Space Shape and Position

TABLE CONTENTS

CHAPTER 1 - 2D SHAPES

CHAPTER 2 - POLYGONS

CHAPTER 3 - OPEN, CLOSED AND PLANE SHAPES

CHAPTER 4 - REGULAR AND IRREGULAR POLYGONS

CHAPTER 5 - PROPERTIES OF QUADRILATERALS

CHAPTER 6 - SYMMETRY

CHAPTER 7 - TESSELLATION

CHAPTER 8 - 3D SHAPES

CHAPTER 9 - NETS

ISBN-13: 978-1505358551
ISBN-10: 1505358558

TWO DIMENSIONAL (2D)

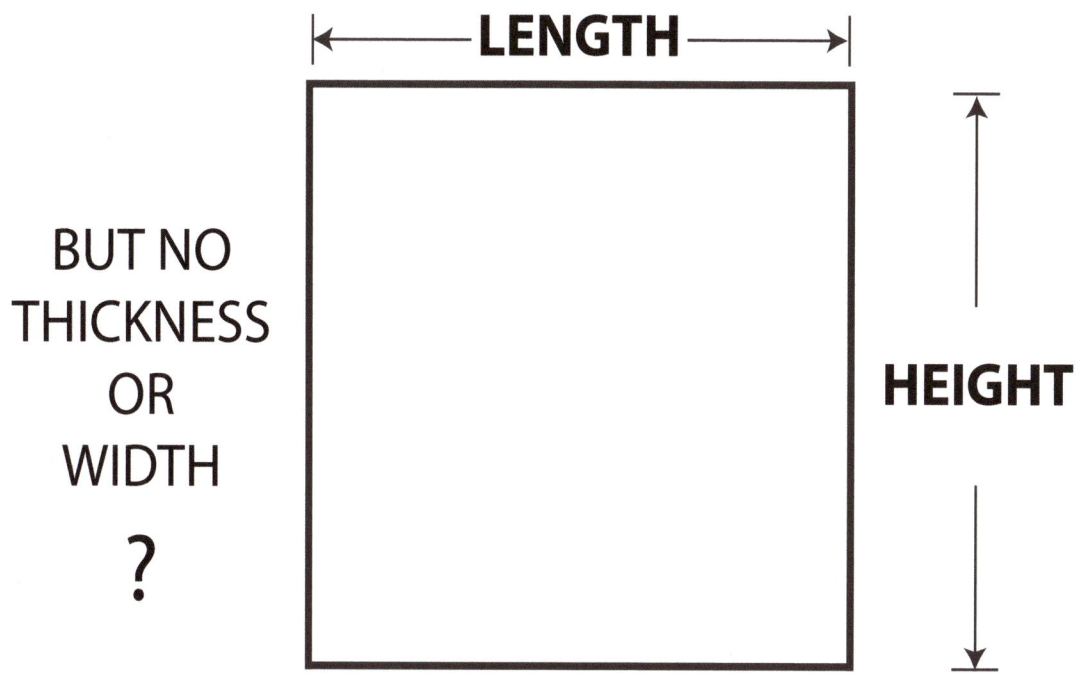

BUT NO
THICKNESS
OR
WIDTH

?

LENGTH

HEIGHT

2D

2 D is short for
TWO DIMENSIONAL
This means that 2D shapes have only
LENGTH and **HEIGHT** but **NO WIDTH (Thickness)**

Is a this square a 2D shape?

THREE DIMENSIONAL (3D)

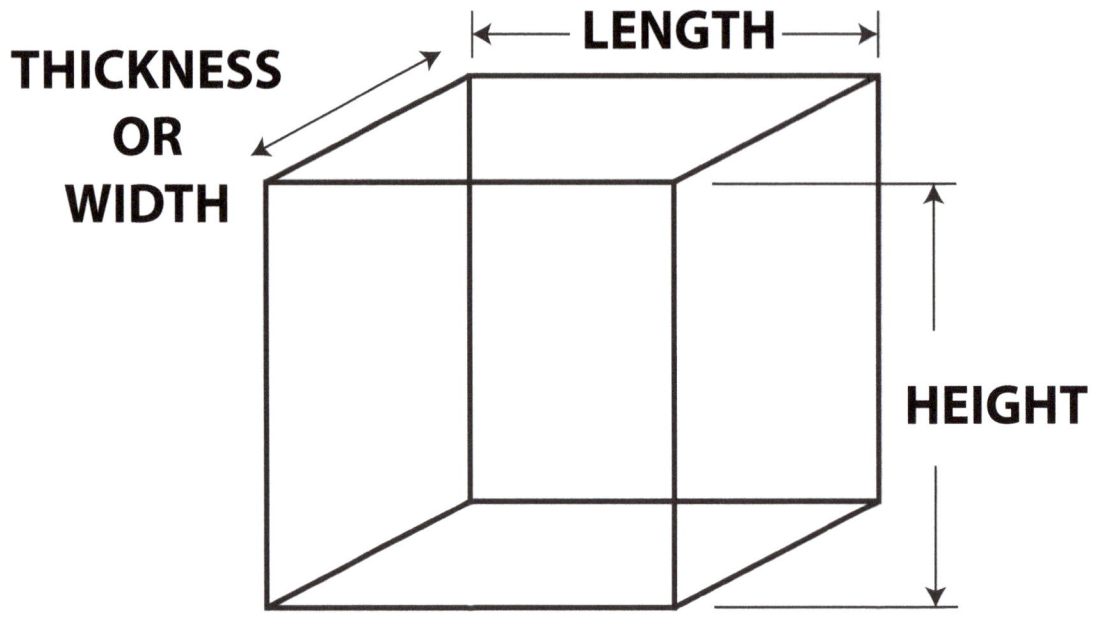

3D

3 D is short for
THREE DIMENSIONAL
This means that 3D shapes have
LENGTH and **HEIGHT** and **WIDTH (Thickness)**
It is a SOLID OBJECT because it has these 3 dimensions

Is a this cube a 3D shape? _____

Is this cube a Solid Object? _____

What is a side?...

A **side** is a straight line that makes part of the shape.

What's an corner?...

A **corner** is where two sides meet.

What's an angle?...

An **angle** is formed when two lines go in different directions from the same point.

How many sides does a square have?

How many corner's does a square have?

How many angle's does a square have?

What is a Polygon?...

A **Polygon** is a closed 2D shape with **3 or more sides** of straight lines.

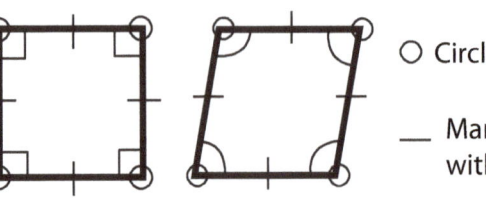

○ Circle the corners

— Mark the sides with a line

Draw in the angle.

If its two straight lines meeting each other, do it like this: ⌐

If the lines are not straight, draw them like this: (

How many sides, corners and angles do all these Polygons have? Fill in the answers underneath each shape.

These are all also **Quadrilateral Polygons** because they all have **4** sides.

Square

Sides _____
Corners _____
Angles _____

Rectangle

Sides _____
Corners _____
Angles _____

Trapezium

Sides _____
Corners _____
Angles _____

Rhombus

Sides _____
Corners _____
Angles _____

Parallelogram

Sides _____
Corners _____
Angles _____

Kite

Sides _____
Corners _____
Angles _____

Properties of Polygons

Counting sides, corners and angles of Polygons.

○ Circle the corners

— Mark the sides with a line

Draw in the angle.

If its two straight lines meeting each other, do it like this: ⌐

If the lines are not straight, draw them like this: ⌒

How many sides, corners and angles do all these Polygons have? Fill in the answers underneath each shape.

Triangle

Sides _____
Corners _____
Angles _____

Pentagon

Sides _____
Corners _____
Angles _____

Hexagon

Sides _____
Corners _____
Angles _____

Heptagon

Sides _____
Corners _____
Angles _____

Octagon

Sides _____
Corners _____
Angles _____

Nonagon

Sides _____
Corners _____
Angles _____

Decagon

Sides _____
Corners _____
Angles _____

Dodecagon

Sides _____
Corners _____
Angles _____

Answer these questions about Polygons...

Are **Polygons** 2D or 3D ⟡ _____

*Don't forget that **2D** means length and height with no thickness.*

Which **Polygons** are Quadrilaterals ⟡

1. _____ 2. _____ 3. _____

4. _____ 5. _____ 5. _____

*Don't forget that **Quadrilaterals** always have four sides. 'Quad' means 4.*

1. Which **Polygon** has **3** sides ⟡ _____

2. How many sides does a **Pentagon** have ⟡ _____

3. Which **Polygon** has **6** sides ⟡ _____

4. How many sides does a **Heptagon** have ⟡ _____

5. Which **Polygon** has **8** sides ⟡ _____

6. How many sides does a **Decagon** have ⟡ _____

7. Which **Polygon** has **9** sides ⟡ _____

8. How many sides does a **Dodecagon** have ⟡

Use this to help you count all the sides.

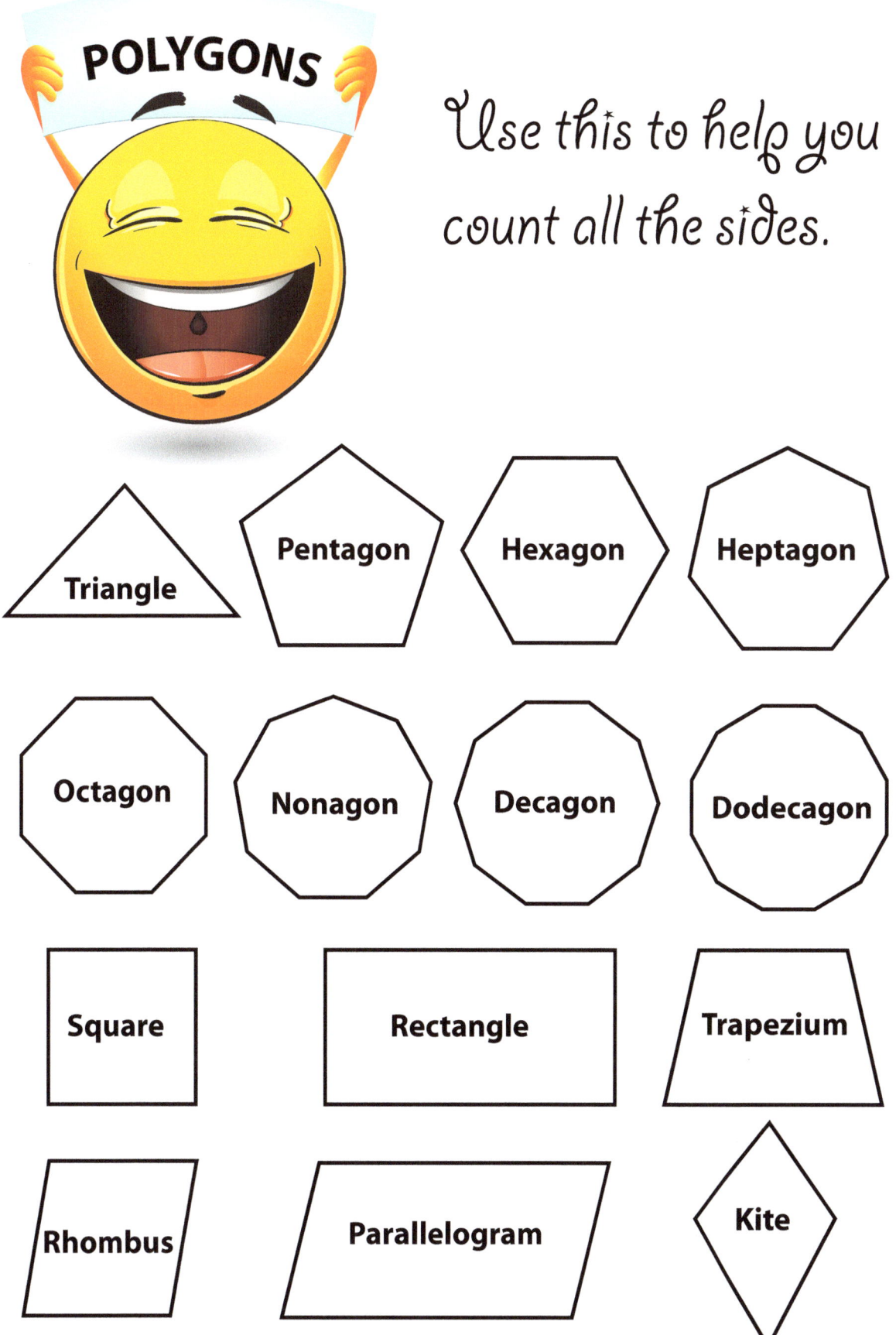

Triangle

Pentagon

Hexagon

Heptagon

Octagon

Nonagon

Decagon

Dodecagon

Square

Rectangle

Trapezium

Rhombus

Parallelogram

Kite

Just one last word about Polygons

 *Don't be confused about Polygons that have been combined into different shapes.*

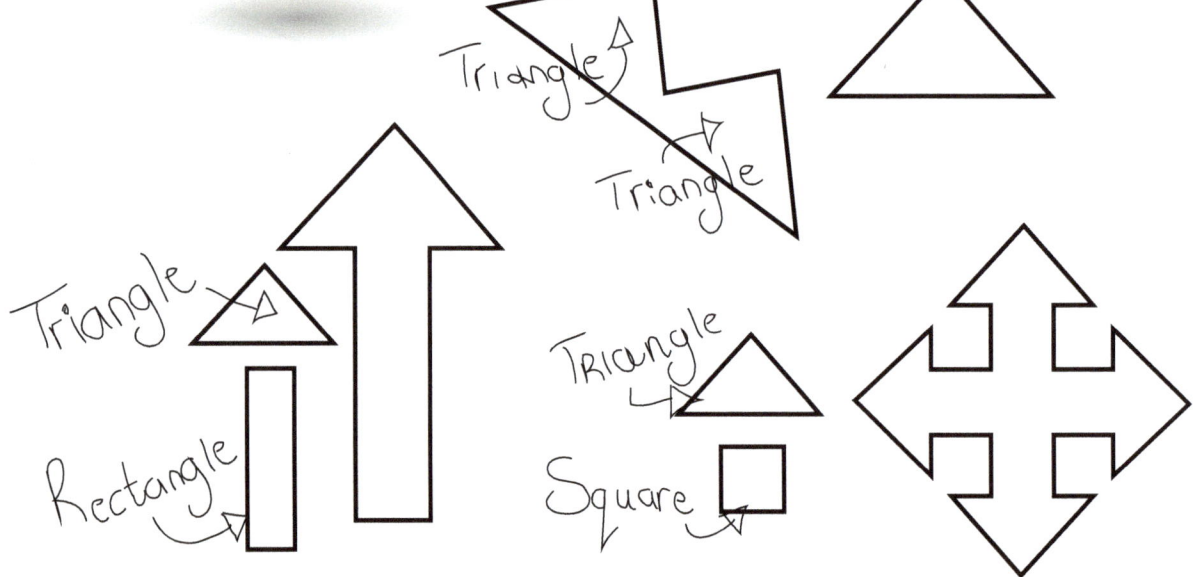

Triangle Triangle

Triangle

Triangle
Rectangle

Triangle
Square

There are different types of triangles.

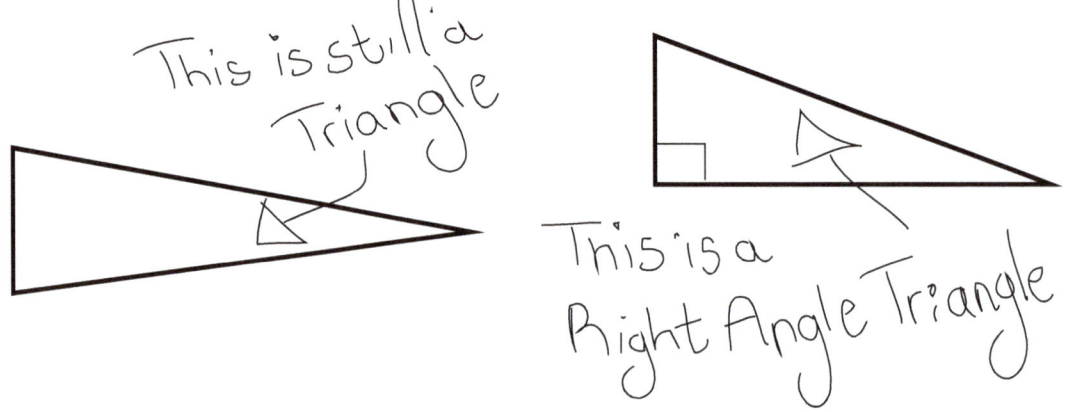

This is still a Triangle

This is a Right Angle Triangle

These two are the same shape

Fill in the Table below:

Shape	Number of sides	Number of angles
Square		
Rectangle		
Triangle		
Circle		
Ellipse		
Pentagon		
Hexagon		
Octagon		
Parallelogram		
Rhombus		
Kite		
Trapezium		

What is an Open Shape?

An **Open** Shape has a different starting point and a different end point.

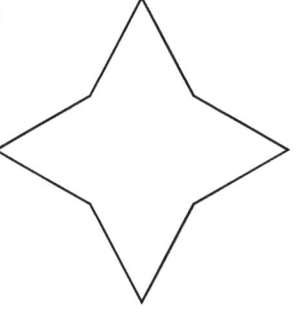

What is a Closed Shape?

A **Closed** Shape has the same start point and end point.

Draw an open shape and a closed shape on the grid below.

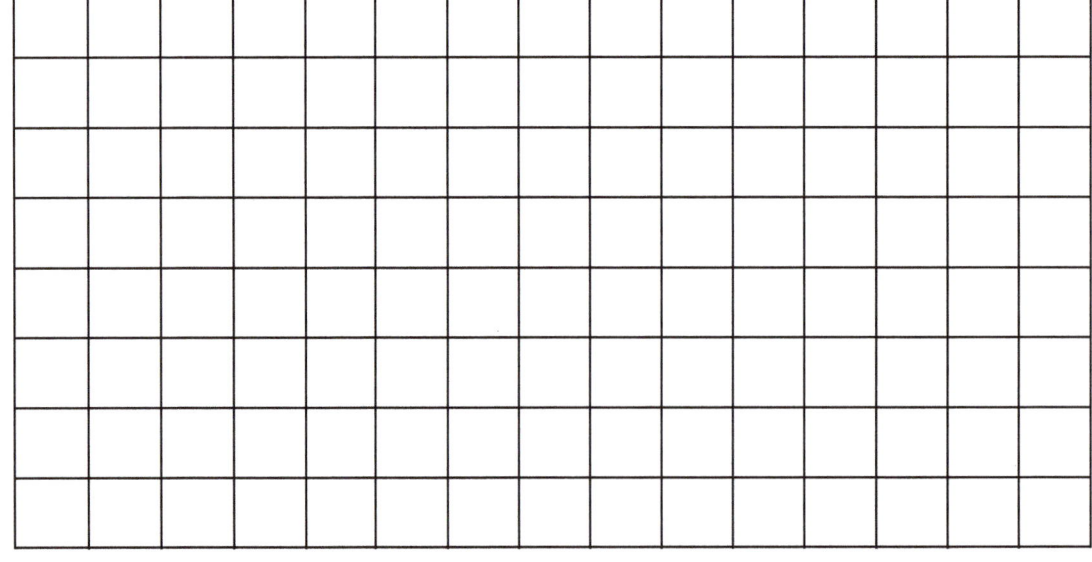

PLANE
FIGURES

What is a Plane figure (shape)?

A **Plane Figure** is a flat figure (2D) with **closed lines** that stays on a single plane. The lines of the figure can be straight, curved or a combination.

A 3D shape cannot be a Plane Figure or a Polygon.

Circle all the **Plane Figures**

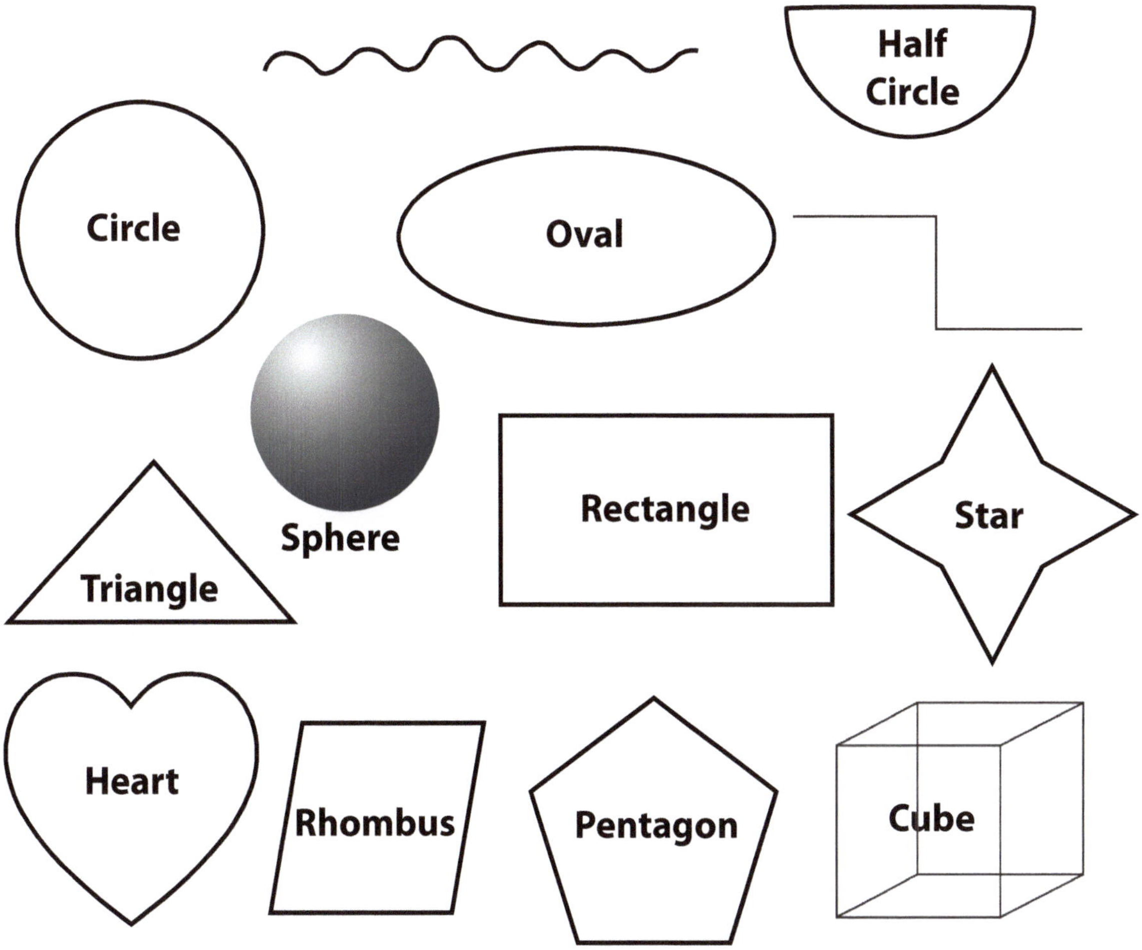

Half Circle

Circle

Oval

Sphere

Rectangle

Star

Triangle

Heart

Rhombus

Pentagon

Cube

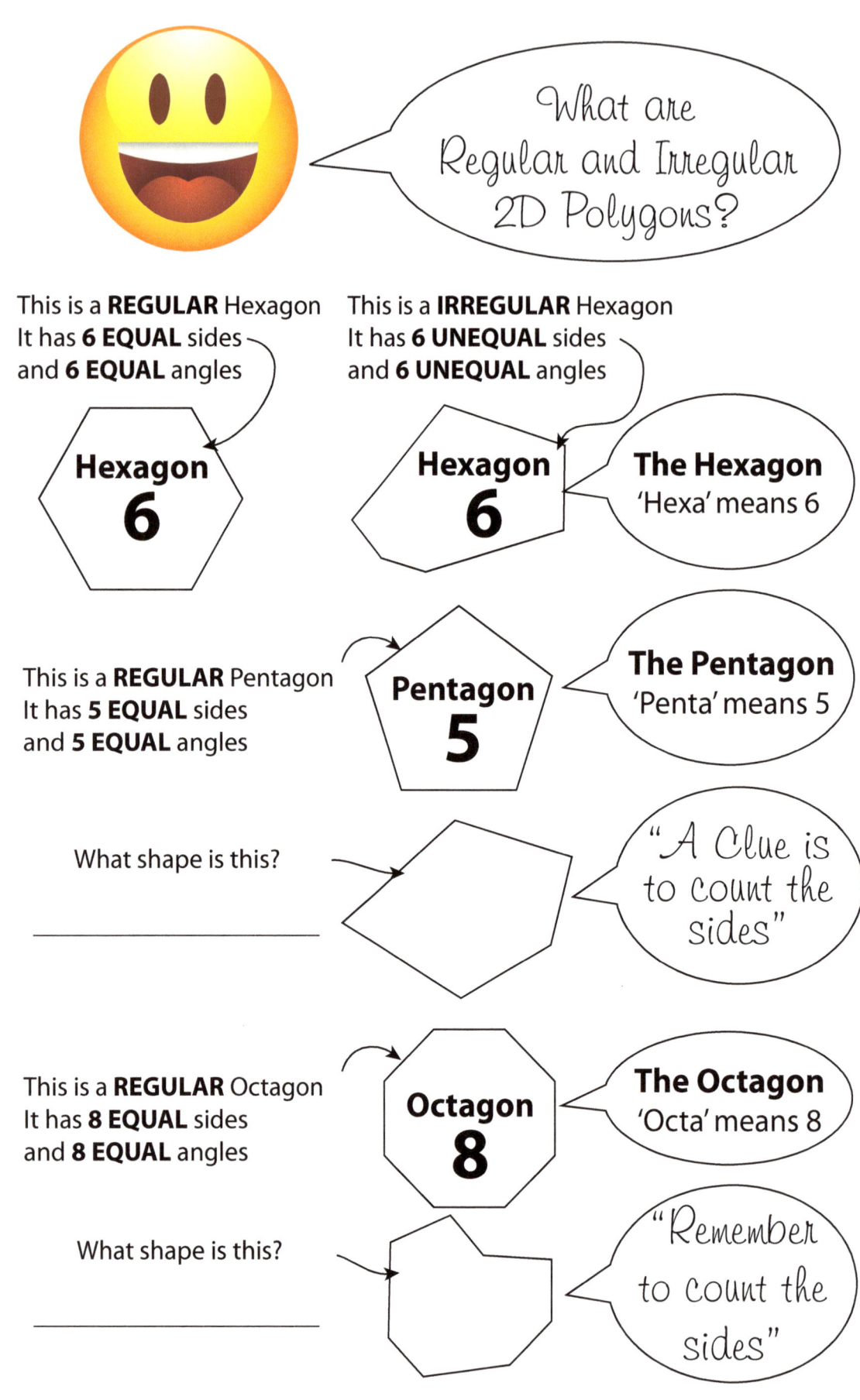

What are Regular and Irregular 2D Polygons?

This is a **REGULAR** Hexagon
It has **6 EQUAL** sides
and **6 EQUAL** angles

Hexagon
6

This is a **IRREGULAR** Hexagon
It has **6 UNEQUAL** sides
and **6 UNEQUAL** angles

Hexagon
6

The Hexagon
'Hexa' means 6

This is a **REGULAR** Pentagon
It has **5 EQUAL** sides
and **5 EQUAL** angles

Pentagon
5

The Pentagon
'Penta' means 5

What shape is this?

_____

"A Clue is to count the sides"

This is a **REGULAR** Octagon
It has **8 EQUAL** sides
and **8 EQUAL** angles

Octagon
8

The Octagon
'Octa' means 8

What shape is this?

_____

"Remember to count the sides"

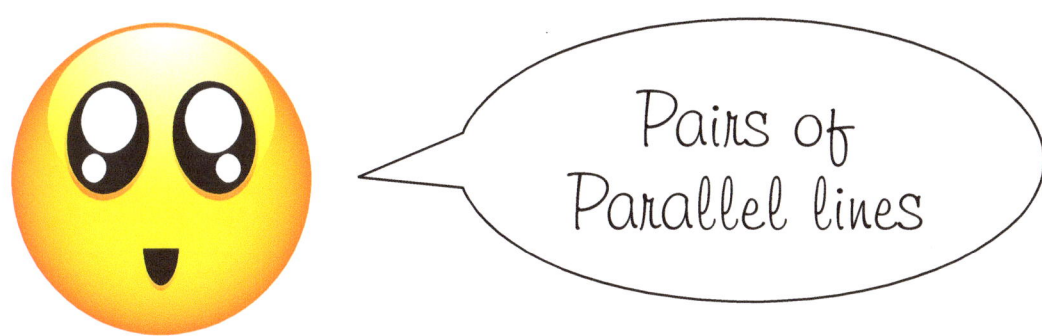

What are Parallel lines?

Parallell lines point in the same direction.

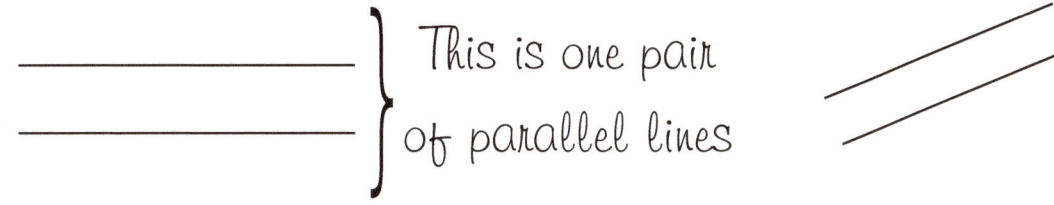

The top line is parallel to the bottom line in both these examples.

How many pairs of parallel lines do all these shapes below have?

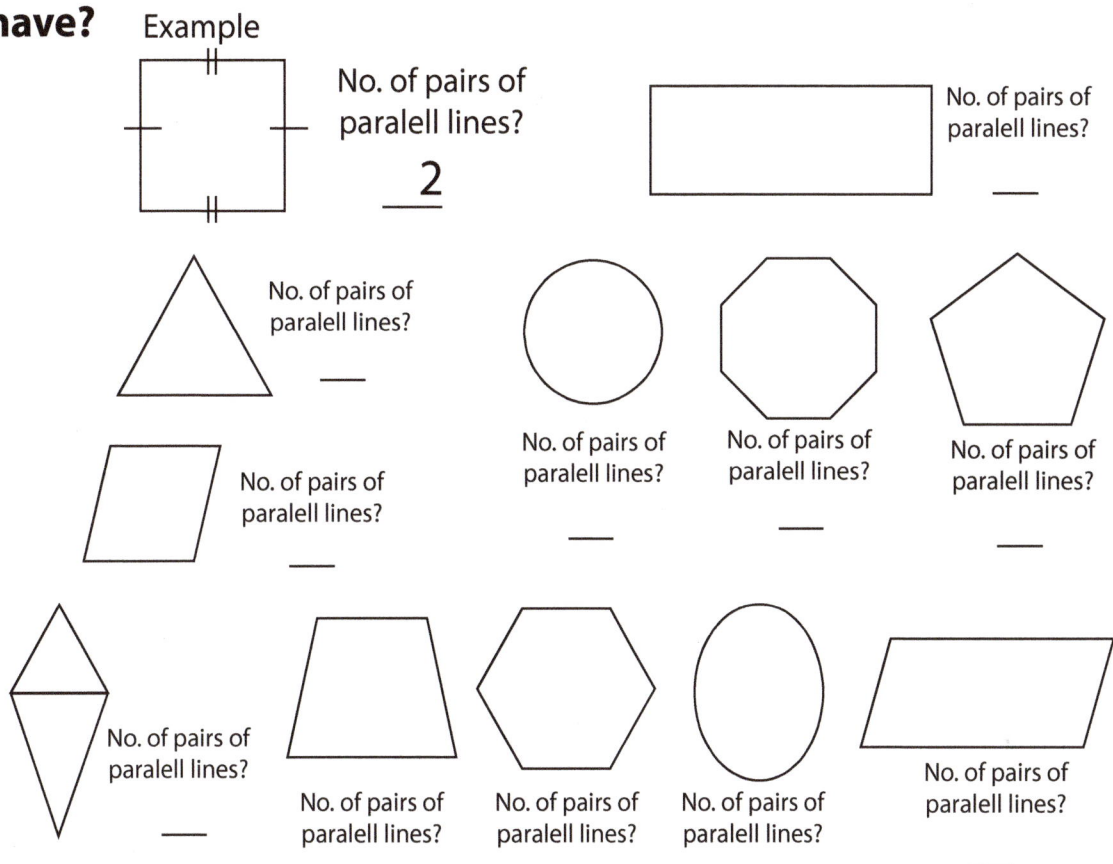

Example

No. of pairs of paralell lines?

_2_

No. of pairs of paralell lines?

___

No. of pairs of paralell lines?

___

No. of pairs of paralell lines?

___

No. of pairs of paralell lines?

___

No. of pairs of paralell lines?

___

No. of pairs of paralell lines?

___

No. of pairs of paralell lines?

___

No. of pairs of paralell lines?

___

No. of pairs of paralell lines?

___

No. of pairs of paralell lines?

___

This is a **SQUARE**

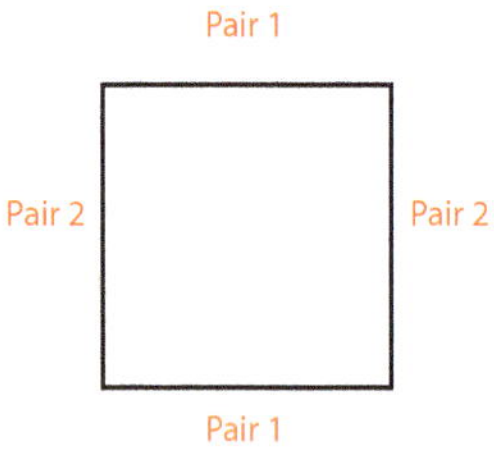

This is a **PARALLELOGRAM**

All 4 sides are the same length

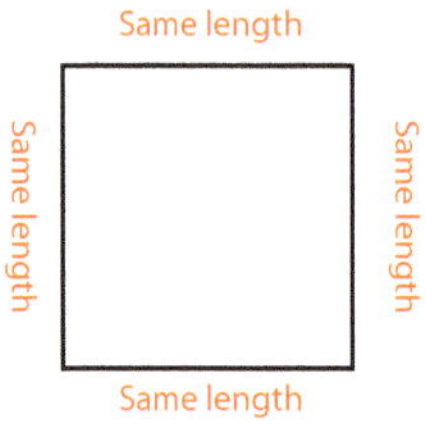

Same length

Same length | Same length

Same length

Only the opposite sides are the same length

Opp. side same length

Opp. side same length | Opp. side same length

Opp. side same length

All angles are 90°

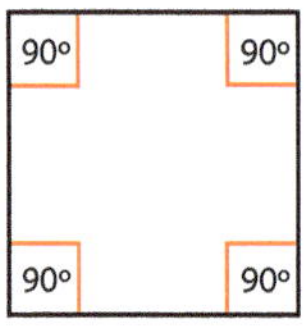

90° | 90°

90° | 90°

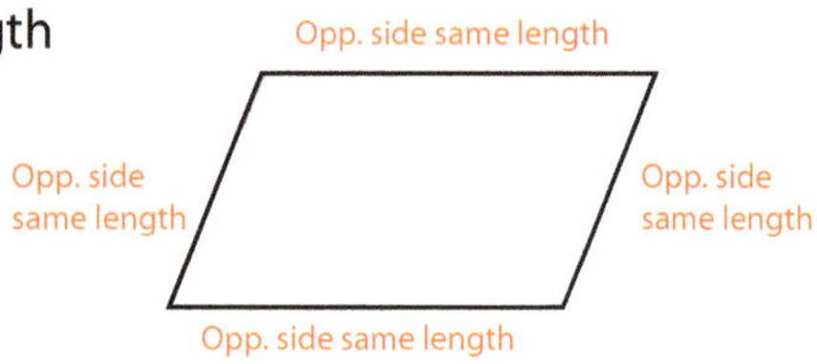

All the opposite angles are equal but not 90°

All diagonals are the same length and cut through each other at opposite angles.

Diagonals are not the same length but do bisect each other at opposite angles.

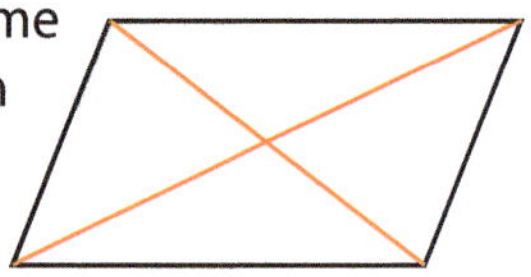

2 pairs of opposite sides which are also parallel.

Pair 1

Pair 2 | Pair 2

Pair 1

2 pairs of opposite sides which are also parallel.

Pair 1

Pair 2 | Pair 2

Pair 1

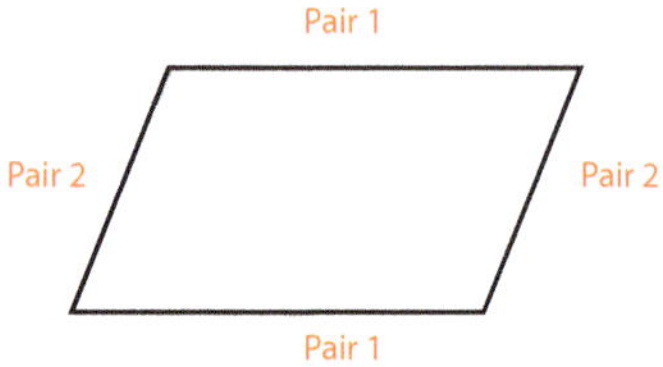

Read carefully all the properties of these quadrilaterals

This is a **RHOMBUS**

This is a **TRAPEZIUM**

All 4 sides are the same length

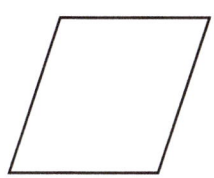

The lines on the side are the same length but the top and bottom are not.

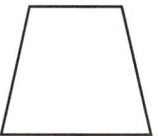

Opposite angles are the same size but are not 90º

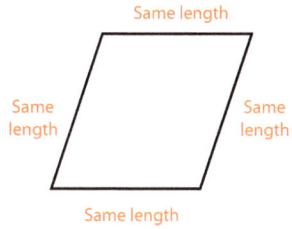

Top two angles are the same and the bottom two angles are the same but not 90º

Diagonals are not the same length but do bisect each other.

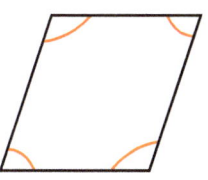

Diagonals are the same length and bisect each other.

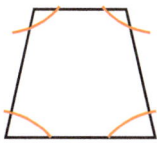

2 pairs of opposite sides which are also parallel.

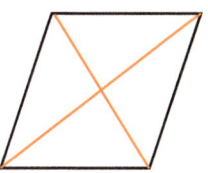

1 pairs of parallel sides.

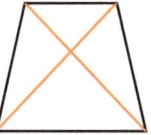

This is a **KITE**

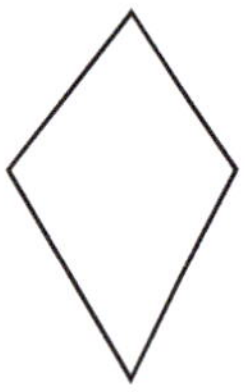

This is a **RECTANGLE**

Top 2 sides equal length
Bottom 2 sides equal length

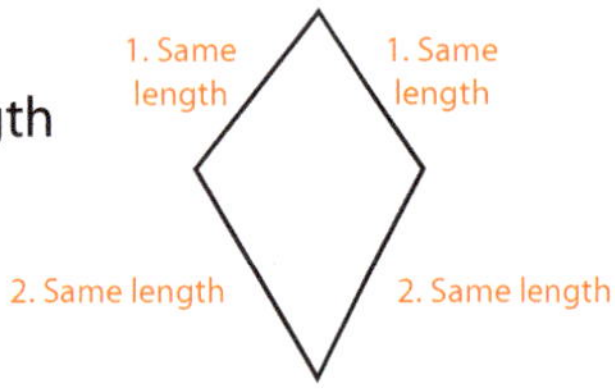

1. Same length   1. Same length
2. Same length   2. Same length

Top and bottom lines are equal length and the left and right sides are equal length.

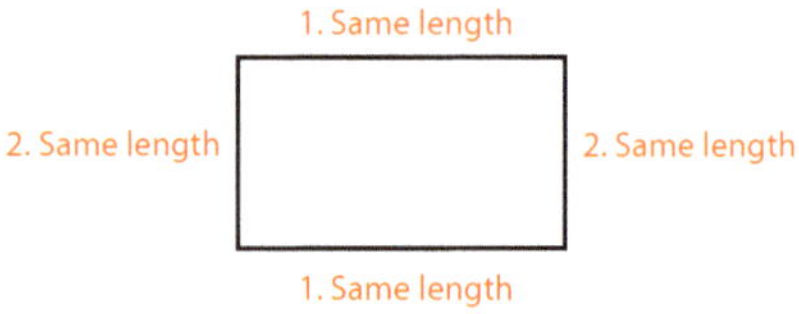

1. Same length
2. Same length   2. Same length
1. Same length

Opposite angles are the same size but are not 90º

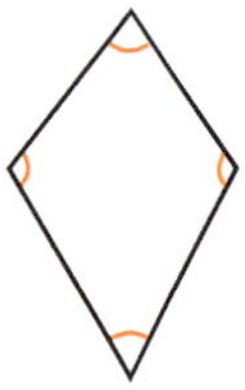

All the angles are 90º

Diagonals are not the same length but do bisect each other.

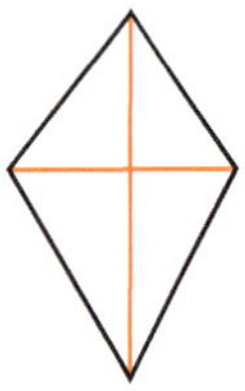

Diagonals are the same length and bisect each other.

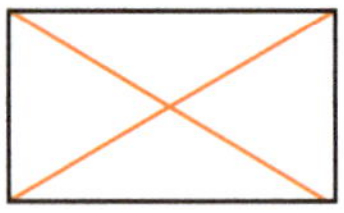

None of the sides are parallel.

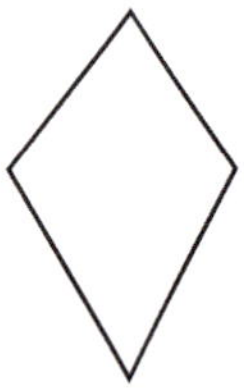

2 pairs of opposite parallel sides.

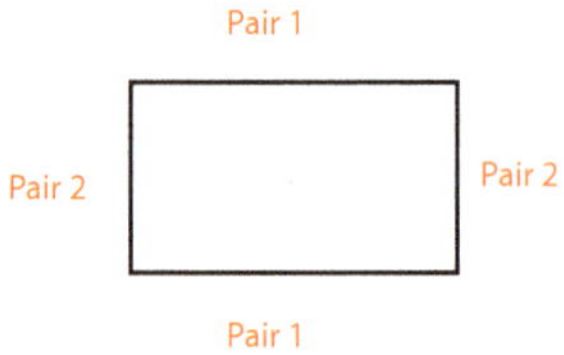

Pair 1
Pair 2   Pair 2
Pair 1

Draw these Quadrilateral Polygons on the grid.

 Take note that a **right angle** is also called a 90º angle and it always looks like a perfect corner, both lines are perfectly straight.

← **right angle (90º)**

What has **4 sides** of **equal length**.

It has **4 equal angles**.

These angles are **right angles**.

The **opposite sides** are **parallell**.

I am a _____

Sometimes I am called **oblong**.

I have **4 sides**.

My **opposite sides** are **equal**

I am a _____

I have **2 pairs** of **equal sides**.

My **opposite sides** are equal in length.

My **opposite angles** are **equal**.

None of my angles are 90º

I am a _____

Sometimes I am known as a **trapezoid**.

I have **one pair** of opposite parallel sides.

I am a _____

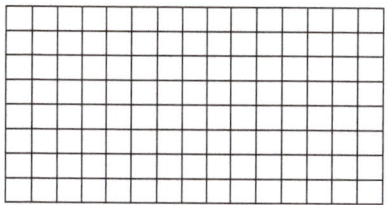

What has **4 sides** of **equal length**.
Opposite angles are not **right angles(90º)**
Opposite angles are **equal (same size)**.
2 pairs of opposite parallel sides.

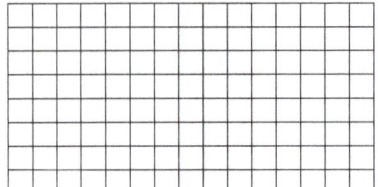

I am a _____

I have **2 pairs** of adjacent sides.
 (does not mean parallell just adjoining each other)
My adjacent sides are **equal in length.**
Opposite angles are the **same size.**
Opposite angles are **not 90º**
None of my sides are **paralell** with each other. (Thats a big clue as there is only one quadrilateral like this).

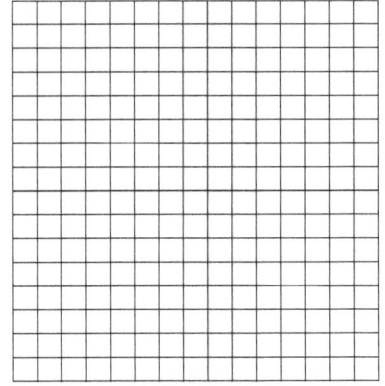

I am a _____

Please circle all the Quadrilateral Polygons on this page.

Remember that a Quadrilateral Polygon has **4 sides** and is also a closed **2D** shape.

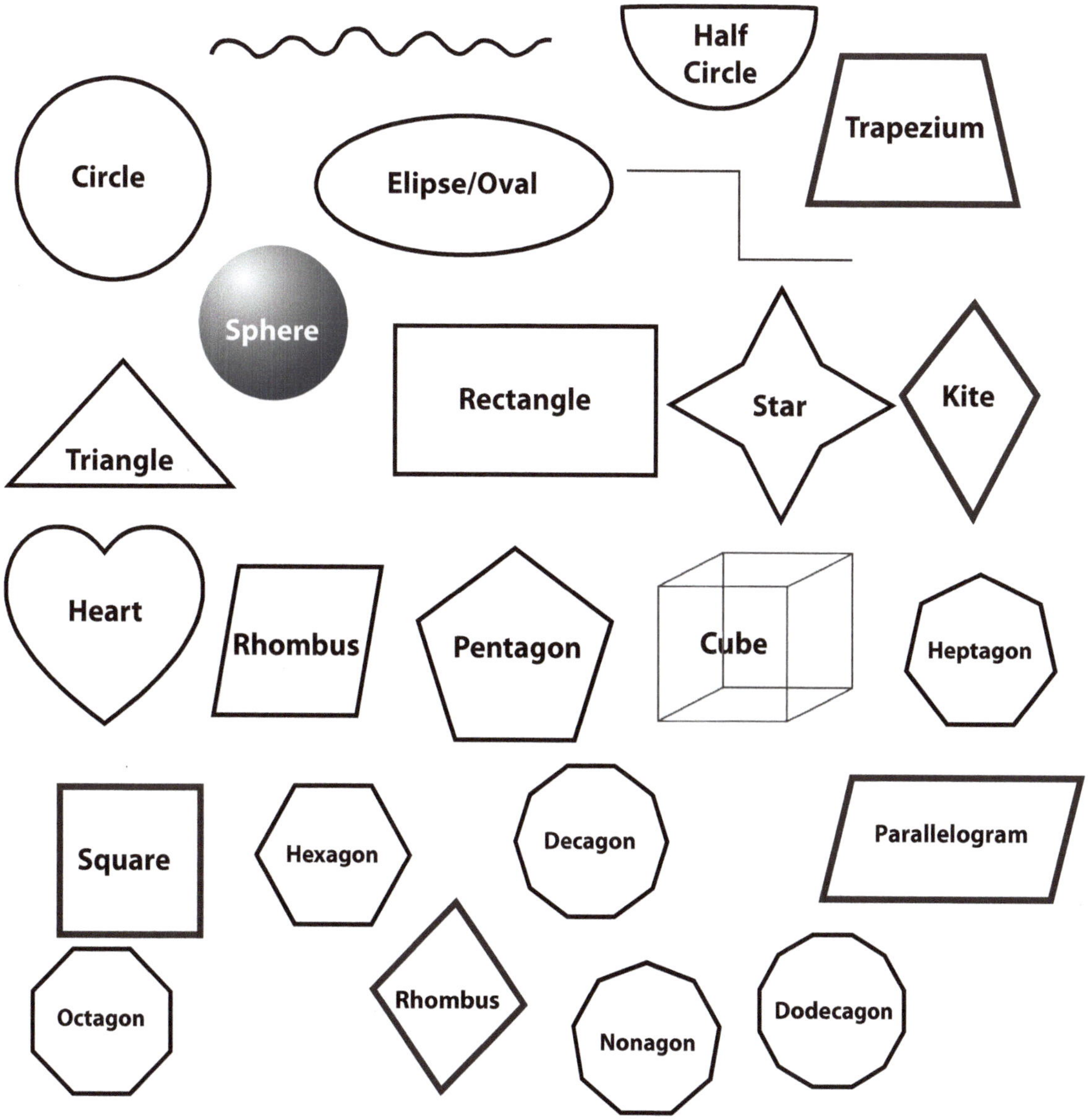

Half Circle

Trapezium

Circle

Elipse/Oval

Sphere

Rectangle

Star

Kite

Triangle

Heart

Rhombus

Pentagon

Cube

Heptagon

Square

Hexagon

Decagon

Parallelogram

Octagon

Rhombus

Nonagon

Dodecagon

Please count how many squares there are in the shapes below.

 The trick is to count the big square first and then all the smaller squares inside.

How many squares are there? _____

What is Symmetry?

Symmetry is when one shape becomes exactly like another when you move it in a certain way (flip, turn or slide it)

So if I take this exact half of the letter 'A' make a copy of it and flip it over, then I will get the other half of the 'A' that looks exactly like the first half. This is a **symmetrical** shape.

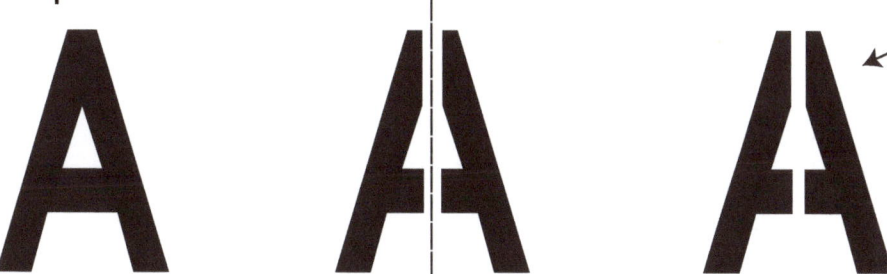

The two halves look the same

Is this 'A' Symmetrical? _____

If I take the letter 'G' and cut it in half make a copy of it and flip it over, I will not get the other half of the 'B'. This means that the letter 'B' is **not symmetrical**

The two halves look different

Is this 'G' Symmetrical? _____

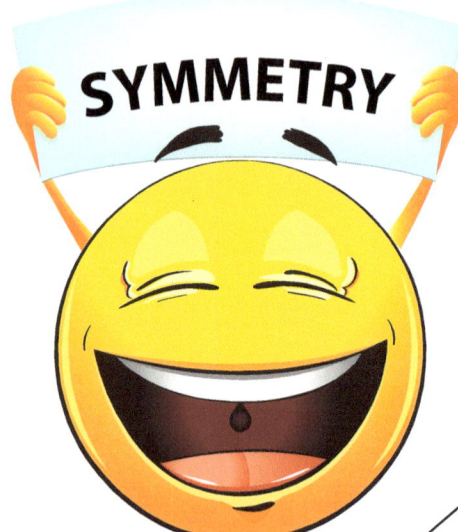

Drawing Symmetry is easy if you take it one square at a time.

The numbers count the blocks to help you.

So here is the line in the middle

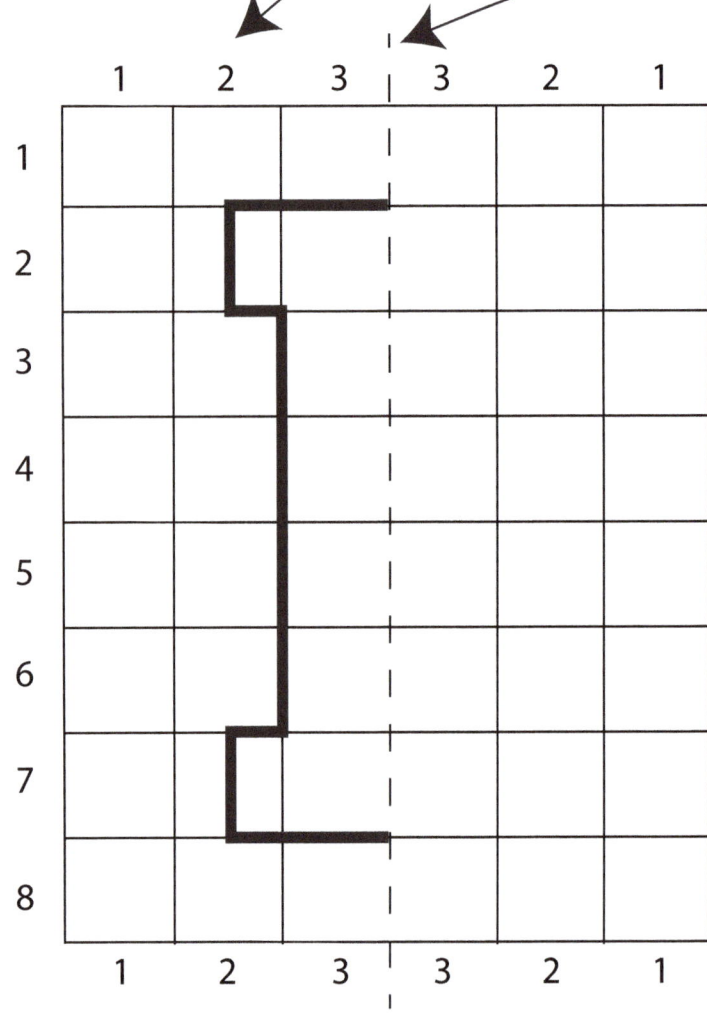

It's easy all you do is count one and a half blocks to the right from the middle line and draw a line up to there.

2 Then draw a line one block down.

3 Then draw a small line half a block towards the left.

4 Then draw one long line down to where the line on the left ends.

6 Draw a small line half a block long to the right and another

7 line one block down.

Then draw a line to the left

8 so that it joins the bottom line on your left.

WELL DONE, YOU DID IT.

SYMMETRY

Let's do this butterfly using the numbered squares to help you.

 Starting from the midde line helps to get it right.

Start from the middle line. Count 4 blocks down and draw a square to the left of the square on the right that is halfway down the block from the top and from the left.

Then draw a line from the top right corner to the bottom left corner.

KEEP GOING LIKE THIS AND YOU WON'T HAVE ANY PROBLEMS DRAWING THIS BUTTERFLY

Start counting from the middle line

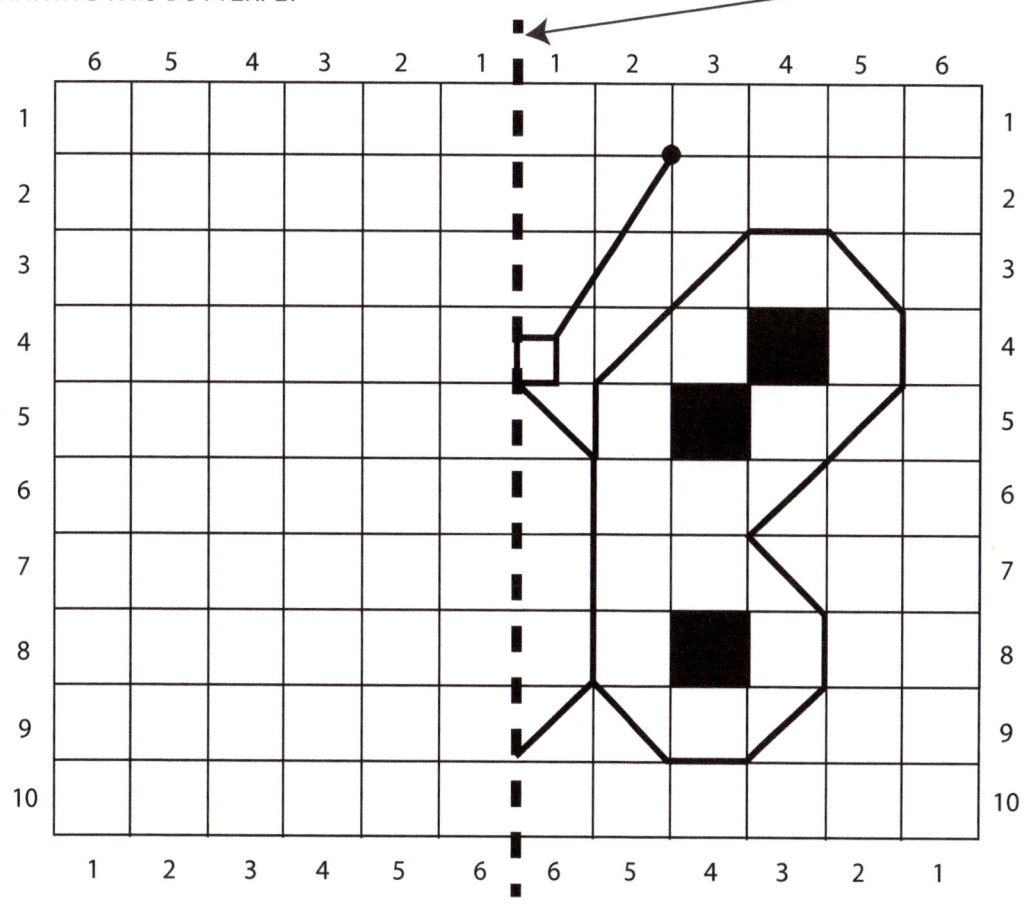

SYMMETRY

You are so cool at this, you can easily do this on your own.

Always count how many block across and how many blocks down before you draw your lines.

Start counting from the middle line

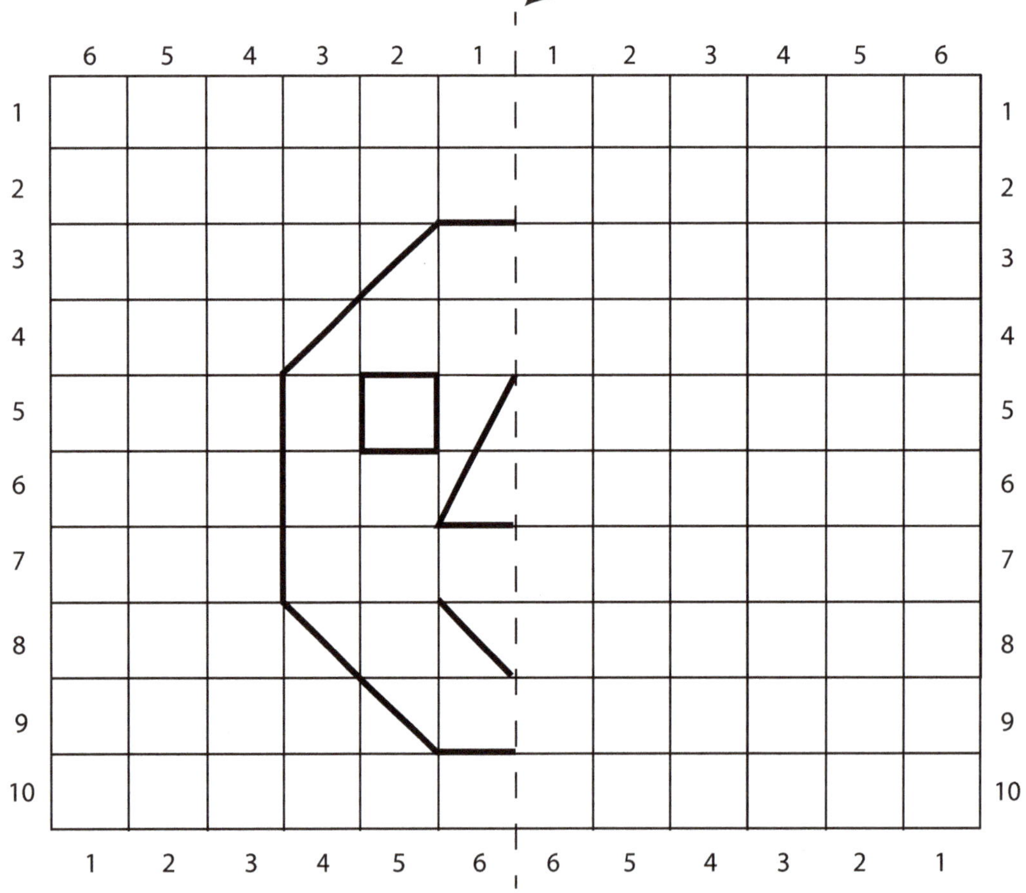

	6	5	4	3	2	1	1	2	3	4	5	6	
1													1
2													2
3													3
4													4
5													5
6													6
7													7
8													8
9													9
10													10

| | 1 | 2 | 3 | 4 | 5 | 6 | 6 | 5 | 4 | 3 | 2 | 1 | |

SYMMETRY

Can you do this

Draw a line from top to bottom.
or from left to right.

Do a line through all these letters of the alphabet and **tick** the ones that have **matching halves**.

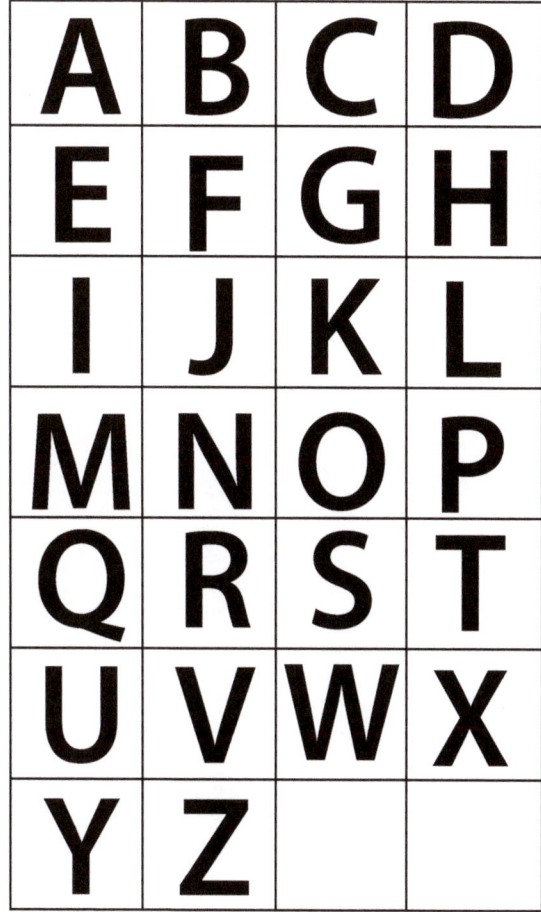

What you are looking for is the lines of symmetry of these letters. Some letters do not have any lines of symmetry.

What exactly is the line or axis of Symmetry

When you fold a shape in half, that fold line is called the line of symmetry. The one half covers the other half exactly.

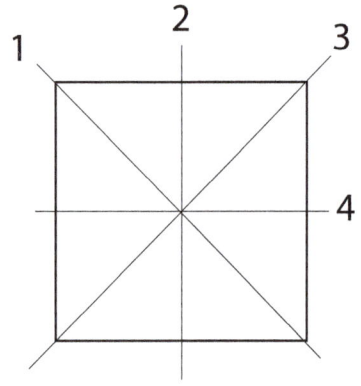

Take the square and fold it in half both ways and then fold it diagonally both ways. If you count each fold line, you will count **4 lines of symmetry.**

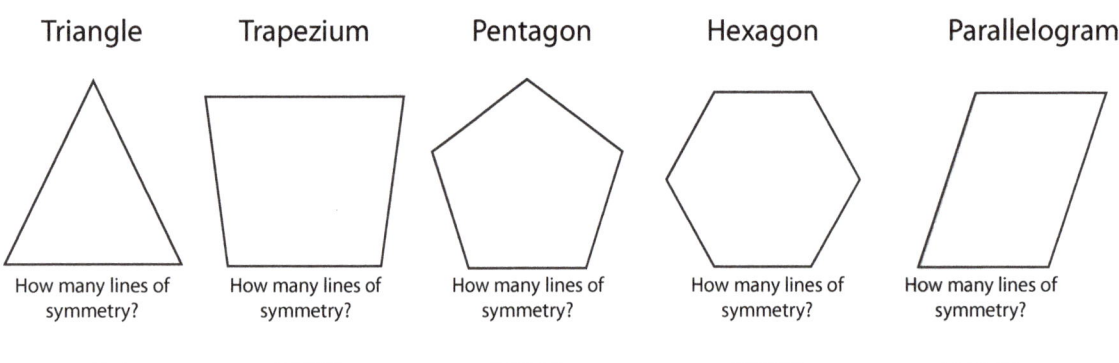

| Triangle | Trapezium | Pentagon | Hexagon | Parallelogram |

How many lines of symmetry? _____
How many lines of symmetry? _____
How many lines of symmetry? _____
How many lines of symmetry? _____
How many lines of symmetry? _____

Take all these shapes and fold them on their lines of symmetry and then draw the lines of symmetry onto the shapes with a ruler and then write underneath each shape how many lines of symmetry you found.

Cut out this shape and fold it on it's line/s of symmetry.

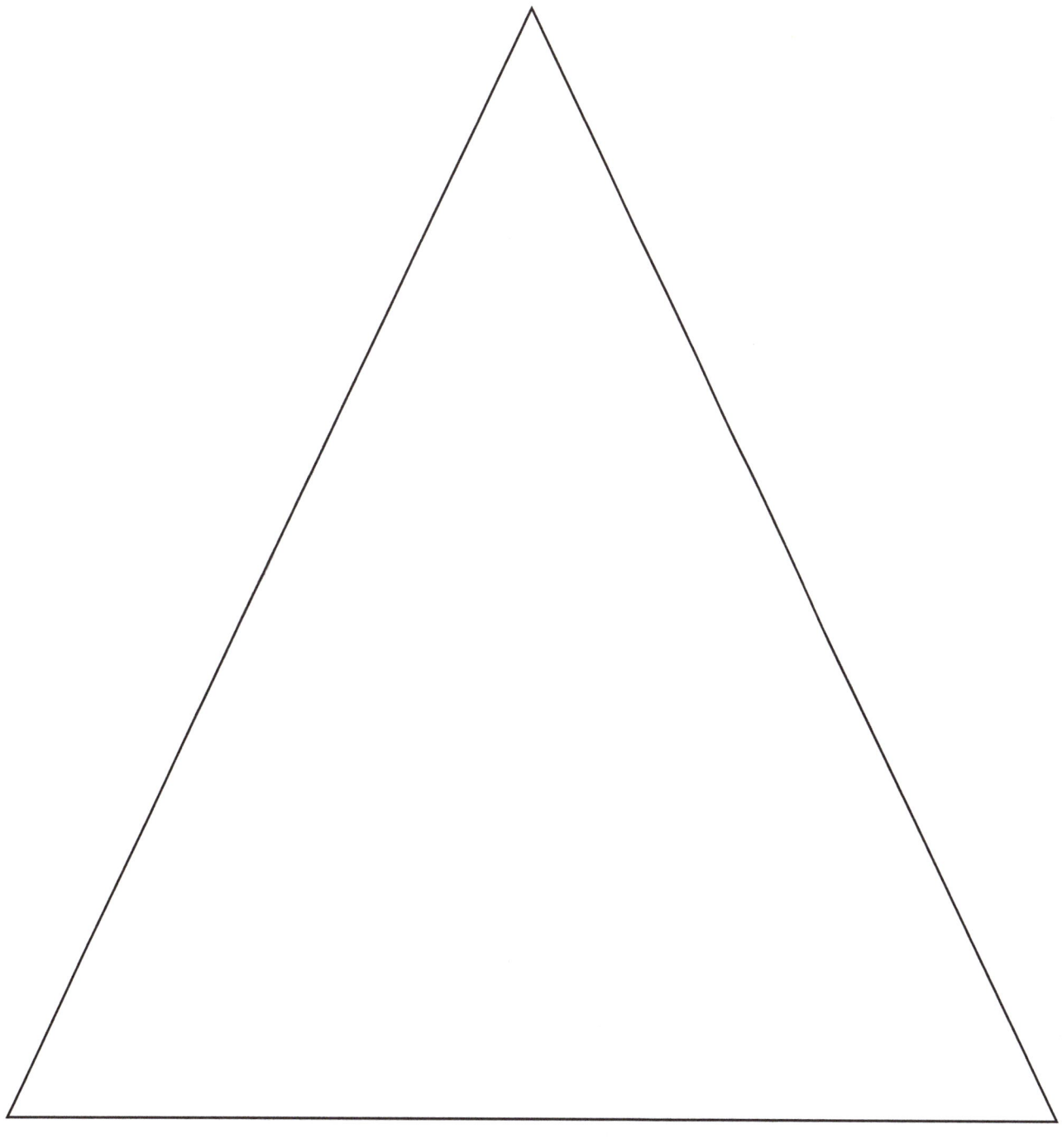

Cut out this shape and fold it on it's line/s of symmetry.

Cut out this shape and fold it on it's line/s of symmetry.

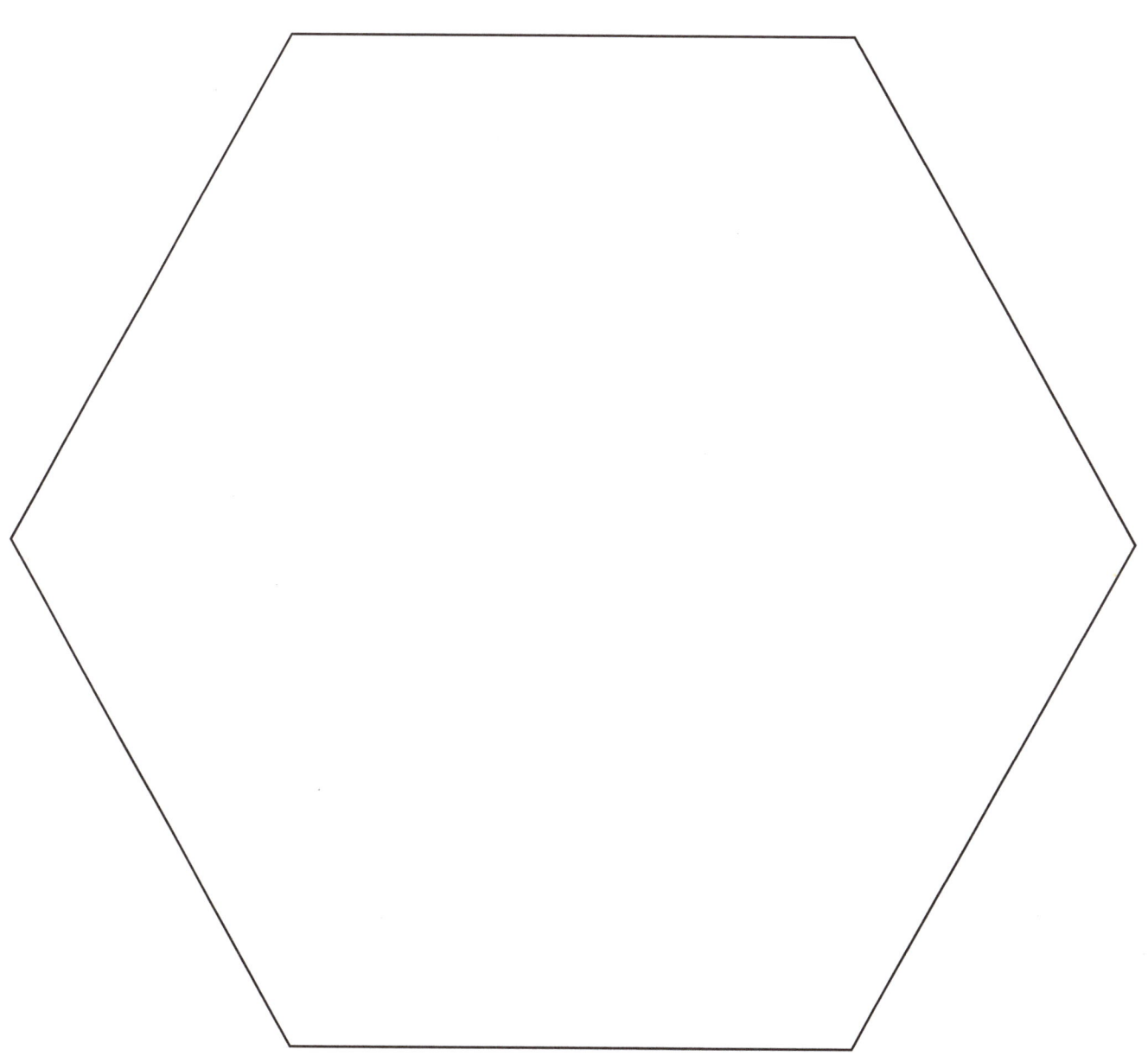

Cut out this shape and fold it on it's line/s of symmetry.

Cut out this shape and fold it on it's line/s of symmetry.

Cut out this shape and fold it on it's line/s of symmetry.

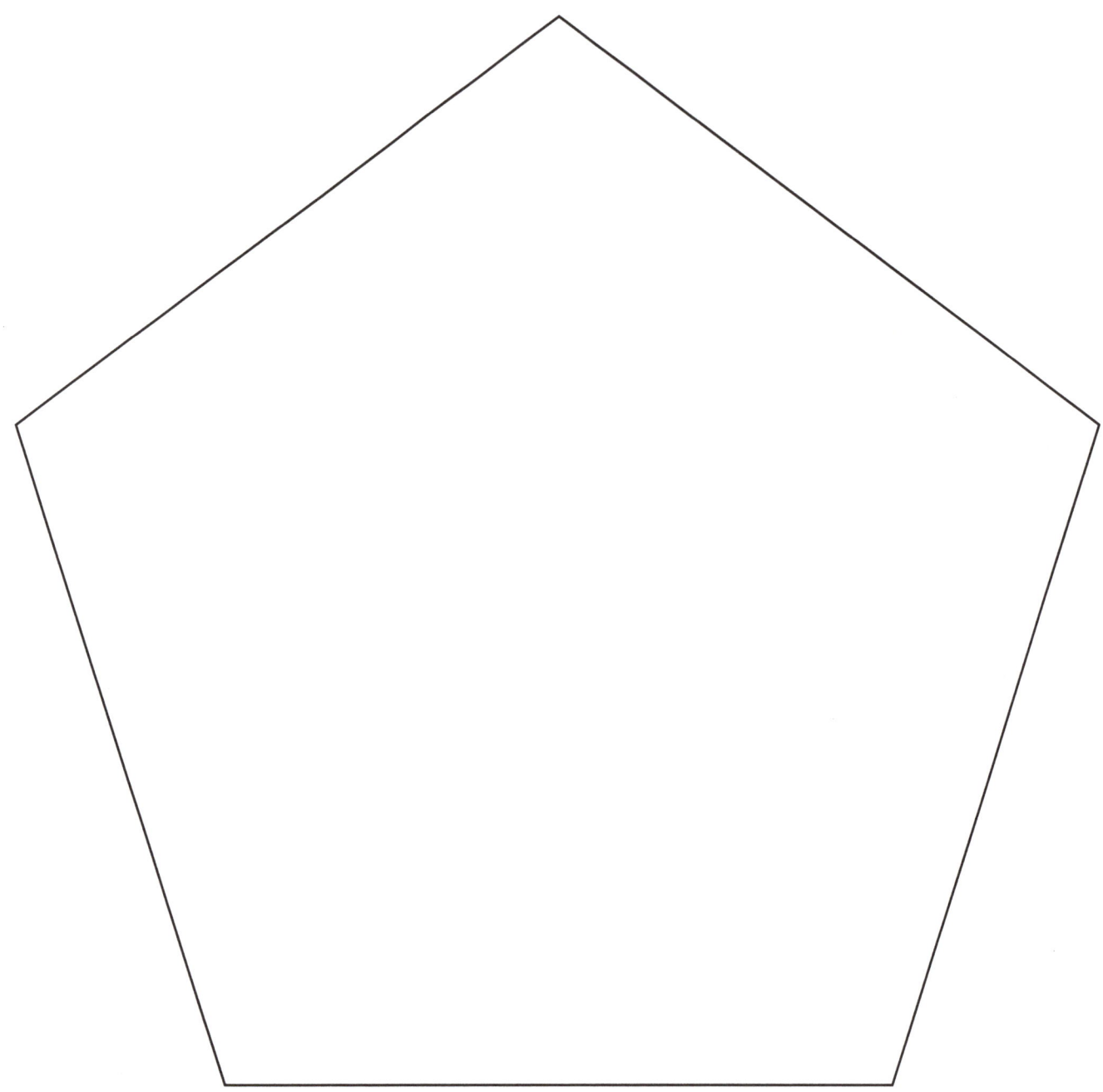

SYMMETRY

Use the line of symmetry and a ruler to complete each shape

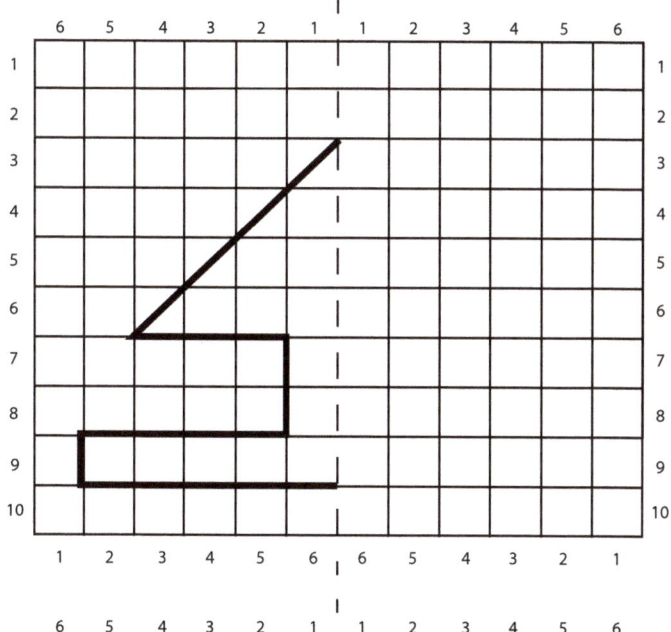

Start at the middle line. Count the blocks that the slanted line takes up 1 2 3 4, then count 1 2 3 4 and draw a line from the top point to 4 squares accross and 4 squares down, in line with the horizontal line.

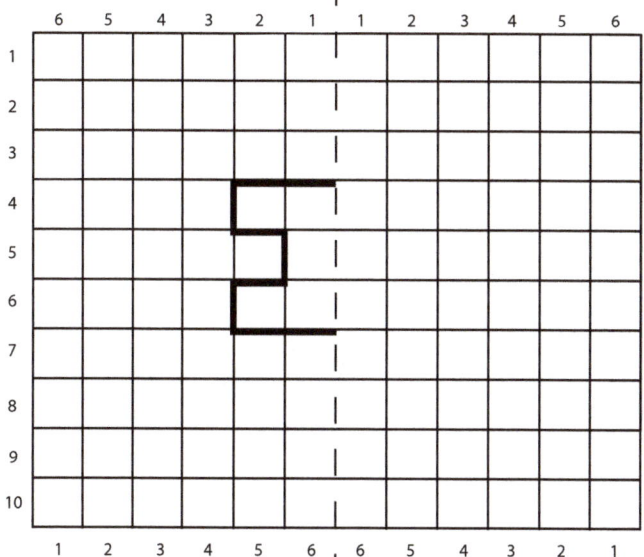

What you end up with is a half that is exactly the same as the half you just copied. It is symmetrical

We know what symmetry is but what is tessellation?

Take these arrows that I have cut out for you and place then one on top of the other. Then flip the one over, then put it back and slide the one and then put it back and turn the one.

Flip   **Slide**   **Turn**

Write underneath each one. Did we Flip it, Slide it or Turn it

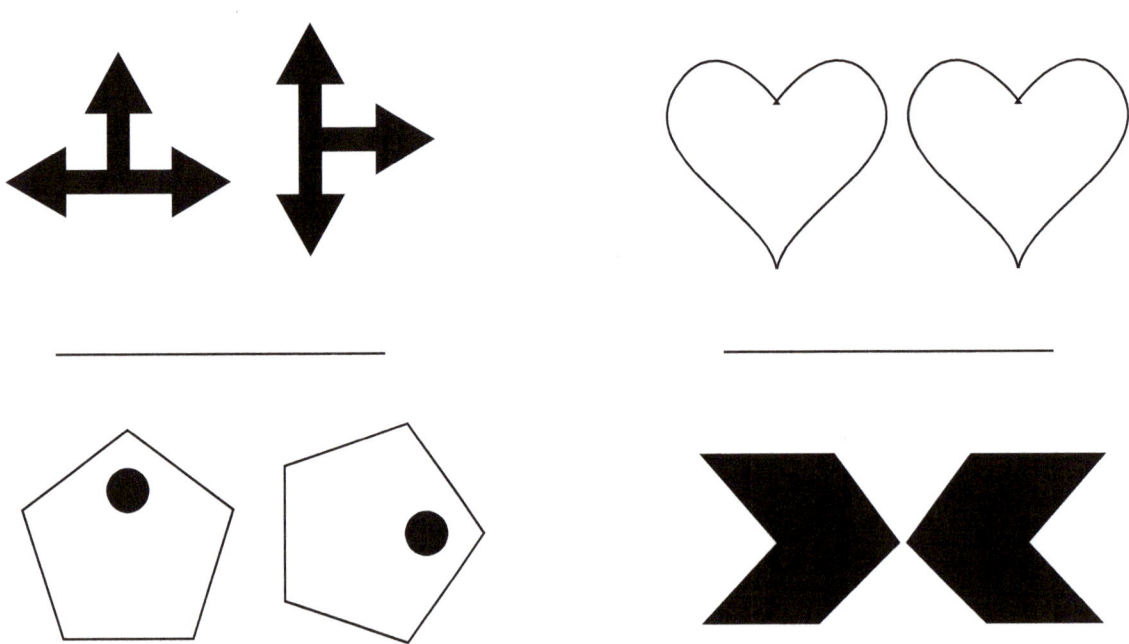

_____   _____

Flip the design in each sqare to make a pattern on the grid

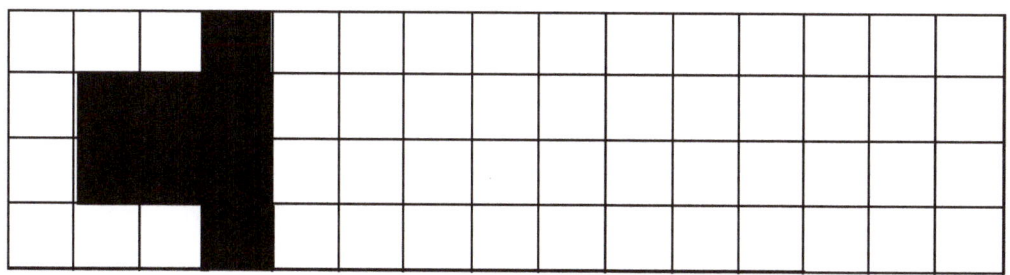

 **Remember what the tessellation flip is!**

Flip

Turn the design in each sqare to make a pattern along the grid

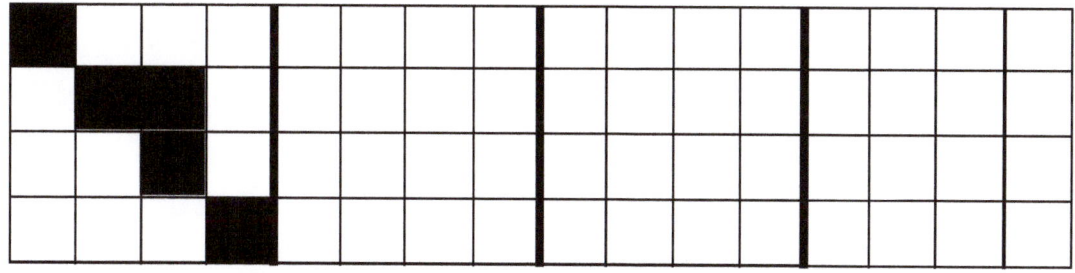

 **Remember what the turn is!**

Turn

THREE DIMENSIONAL (3D)

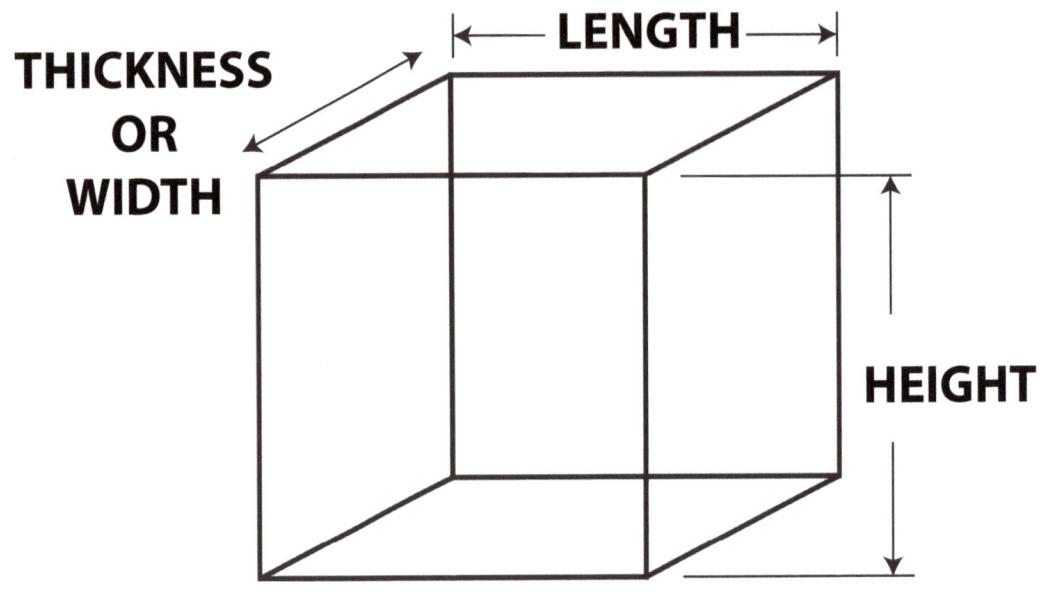

THICKNESS OR WIDTH
LENGTH
HEIGHT

3D

3 D is short for
THREE DIMENSIONAL
This means that 3D shapes have
LENGTH and **HEIGHT** and **WIDTH (Thickness)**
It is a SOLID OBJECT because it has these 3 dimensions

Is a this cube a 3D shape? _____

Is this cube a Solid Object? _____

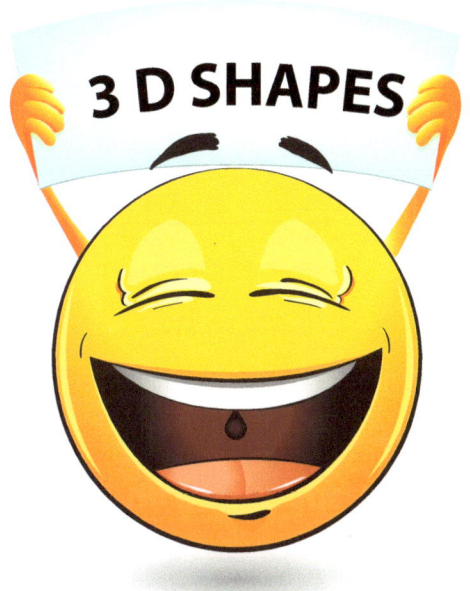

What are faces, corners and edges

Cube

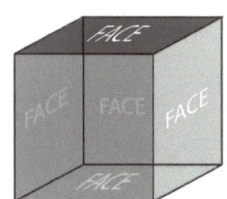

Face
A face is any of the flat surface of a solid object.

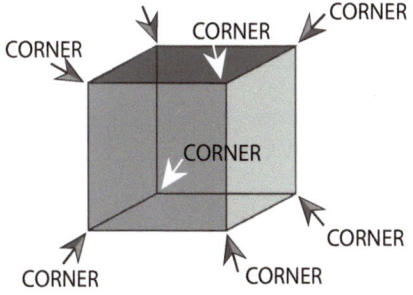

Corner or a vertex
A corner is a point where two or more straight lines meet.

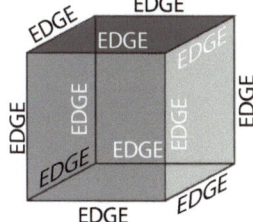

Edge
An edge is the line that joins two corners.

Sphere

No Edge of a sphere or circle
This is called a curved surface.
A sphere has 1 curved surface.

Identifying all your 3D Shapes

Remember that a 3D shape has Length, Height and Thickness. Unlike a 2D shape that has only length and height but no thickness.

Cube

How many faces? _____
How many corners? _____
How many edges? _____
How many curved surfaces? _____

Sphere

How many faces? _____
How many corners? _____
How many edges? _____
How many curved surfaces? _____

Cone

How many faces? _____
How many corners? _____
How many edges? _____
How many curved surfaces? _____

Cylinder

How many faces? _____
How many corners? _____
How many edges? _____
How many curved surfaces? _____

Prisms

 Don't forget to count the front and the back.

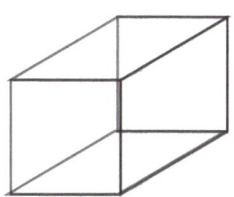

Rectangular Prism

How many square faces? _____
How many rectangular faces? _____
How many corners? _____
How many edges?   _____
How many curved surfaces? _____

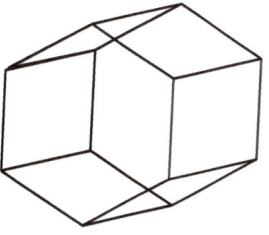

Hexagonal Prism

How many faces? _____
How many corners? _____
How many edges?   _____
How many curved surfaces? _____

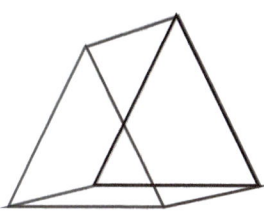

Triangular Prism

How many faces?   _____
How many corners? _____
How many edges?   _____
How many curved surfaces? _____

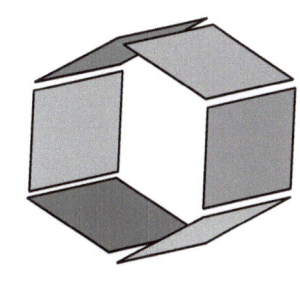

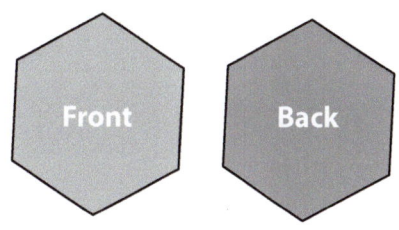

Front   Back

Pyramids

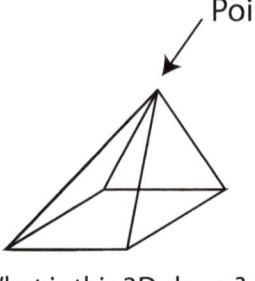

Point

What is this 3D shape?   What is this 3D shape?

_____   _____

The difference is that the first one comes to a point and the second one doesnt!
and what structure in Egypt comes to a point?

Square Based Pyramid

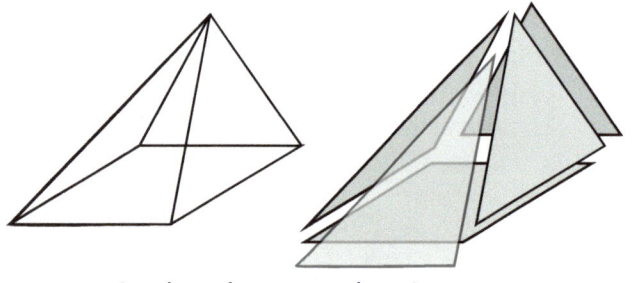

How many faces? _____
How many corners? _____
How many edges? _____
How many curved surfaces? _____

Look at the square base!

Triangular Based Pyramid

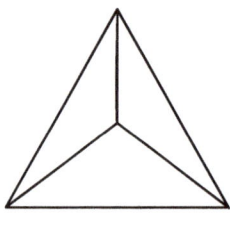 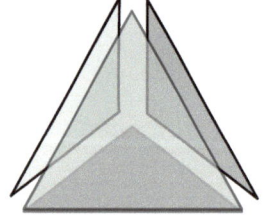

How many faces? _____
How many corners? _____
How many edges? _____
How many curved surfaces? _____

Look at the triangular base!

Can you tell me what all these 3D shapes are?

Remember the shapes that come to a point are pyramids.

What is this 3D shape?

_____

What is this 3D shape?

_____

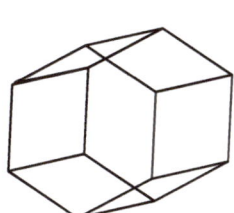

What is this 3D shape?

_____

What is this 3D shape?

_____

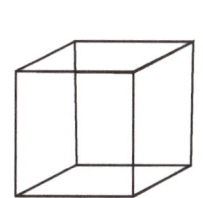

What is this 3D shape?

_____

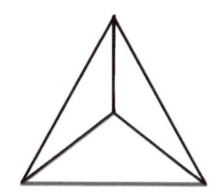

What is this 3D shape?

_____

What is this 3D shape?

_____

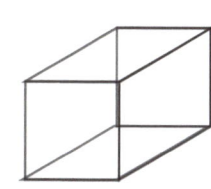

What is this 3D shape?

_____

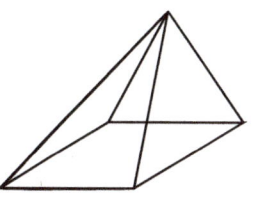

What is this 3D shape?

_____

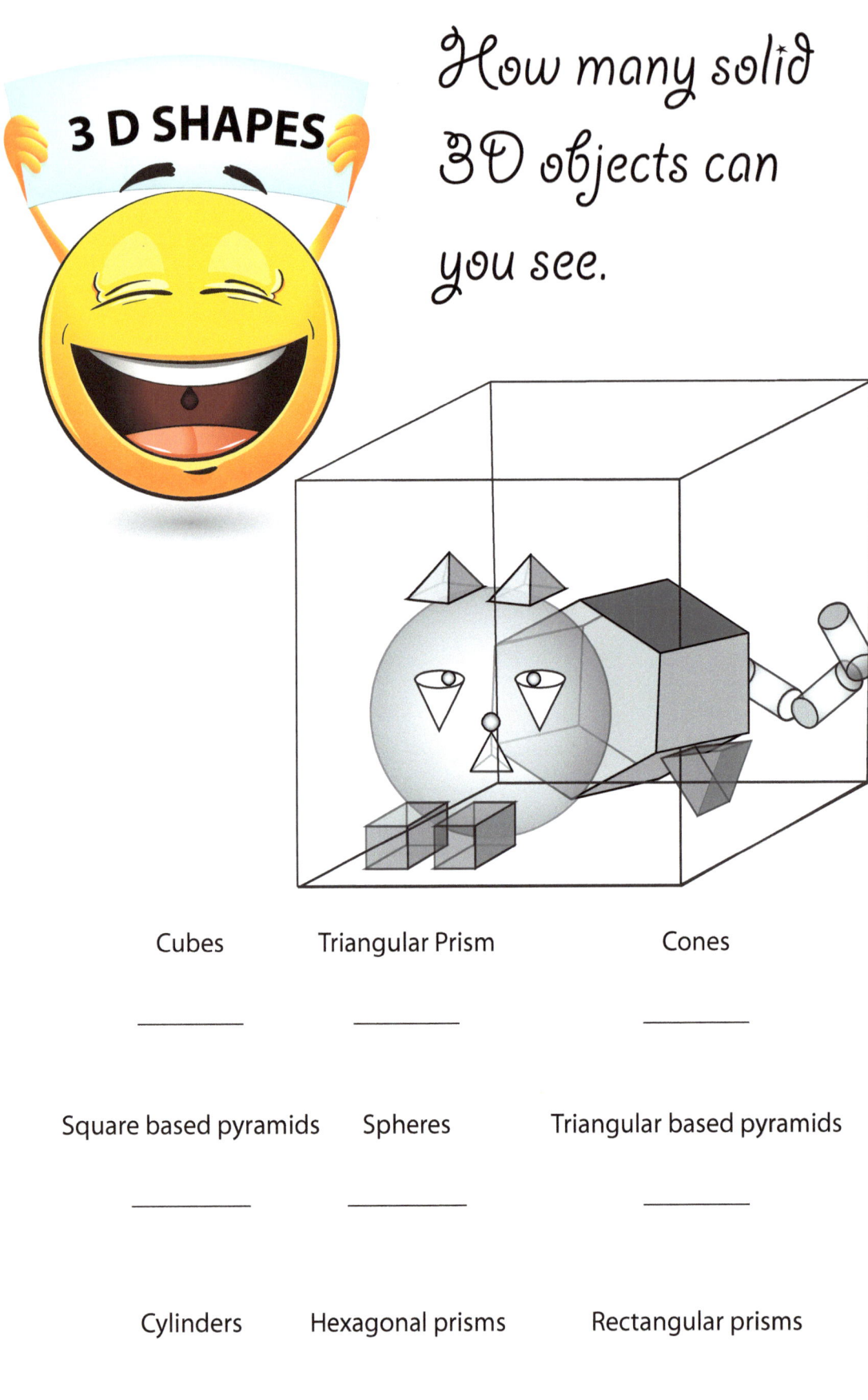

How many solid 3D objects can you see.

Cubes	Triangular Prism	Cones
_____	_____	_____

Square based pyramids	Spheres	Triangular based pyramids
_____	_____	_____

Cylinders	Hexagonal prisms	Rectangular prisms
_____	_____	_____

Its your turn to answer some questions. I know you are going to get 100%

1. Which 3D shape has only one curved surface?_____

2. Which 3D shape has only one face and one curved surface _____

3. Which 3D shape has one curved surface and two faces _____

4. Please cirlcle all the Plane figures.  Remember Plane figures are 2D (flat) and are closed shapes

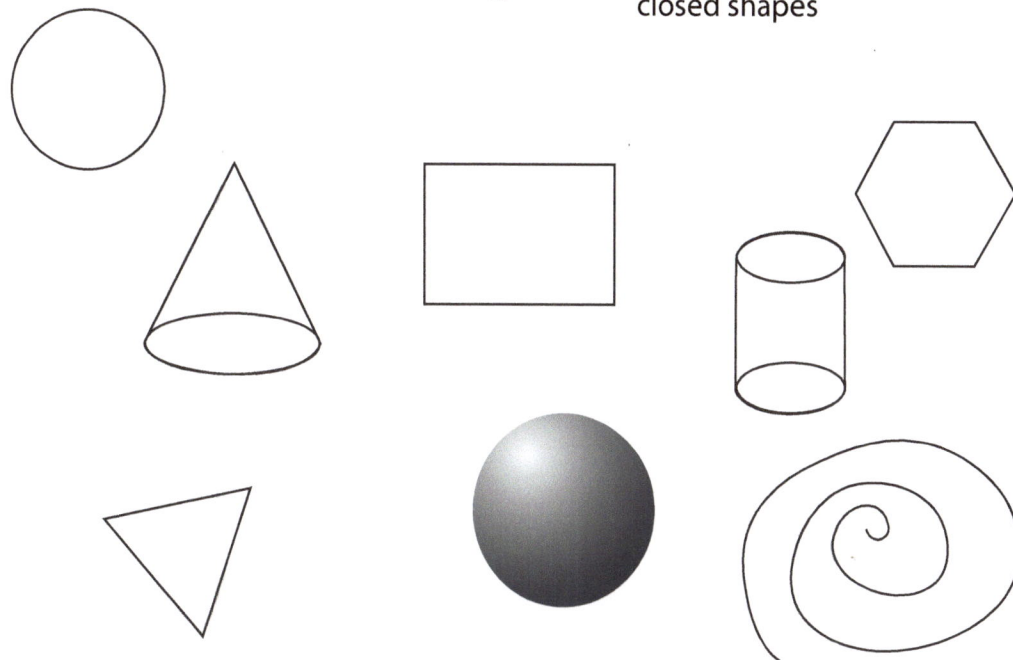

What is a Net?

A pattern that you can cut and fold to make a model of a solid shape.

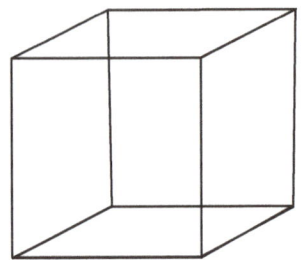

This is a net for a cube.
If you cut it out and fold it you will have a cube.

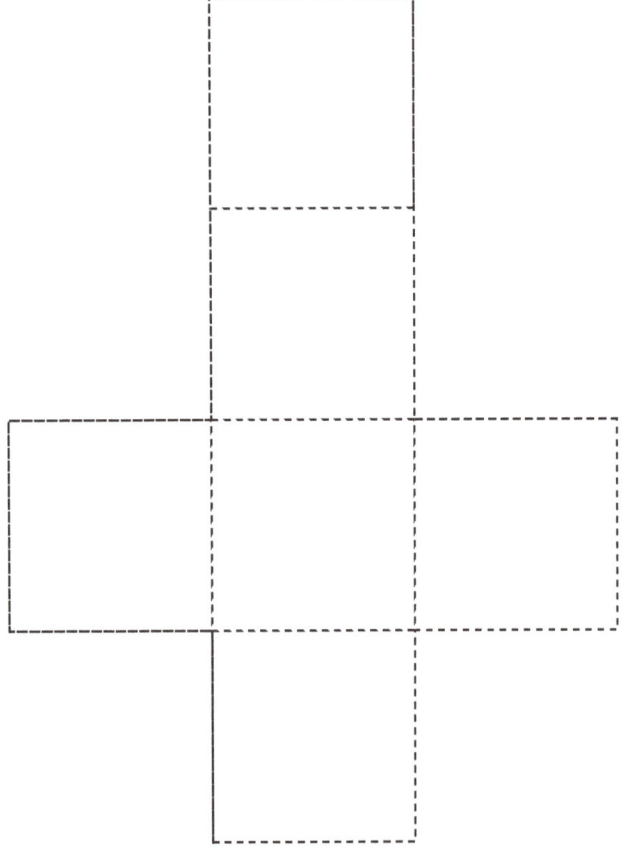

Nets for all the shapes

Im going to show you all the nets that belong to all the different 3D Solid shapes.

Do you see how these nets belong to these shapes.

Triangular Prism

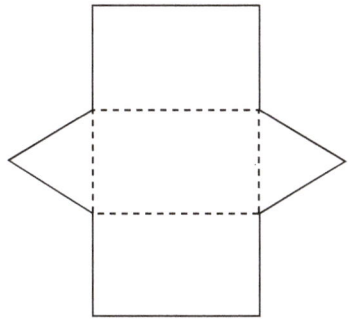

Triangular Prism Net

Cone

Cone Net

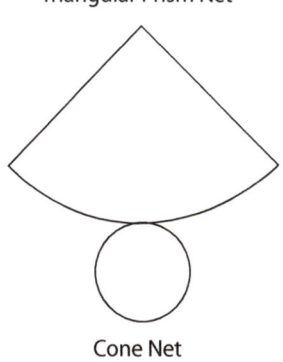

Square Based
Pyramid.

Square Based
Pyramid Net

Nets for all the shapes

Im going to show you all the nets that belong to all the different 3D Solid shapes.

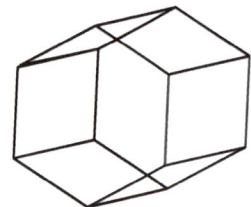

Hexagonal Prism

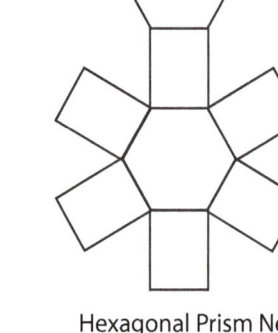

Hexagonal Prism Net

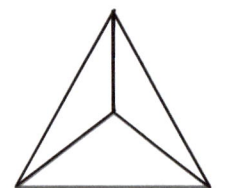

Triangular Based Pyramid

Triangular Based Pyramid Net

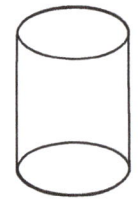

Cylinder

Cylinder Net

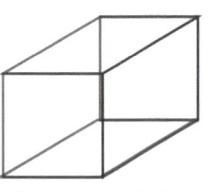

Rectangular Prism

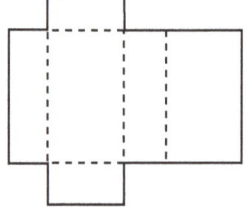

Rectangular Prism Net

Cut out this net and fold it into its 3D shape.

Triangular Prism

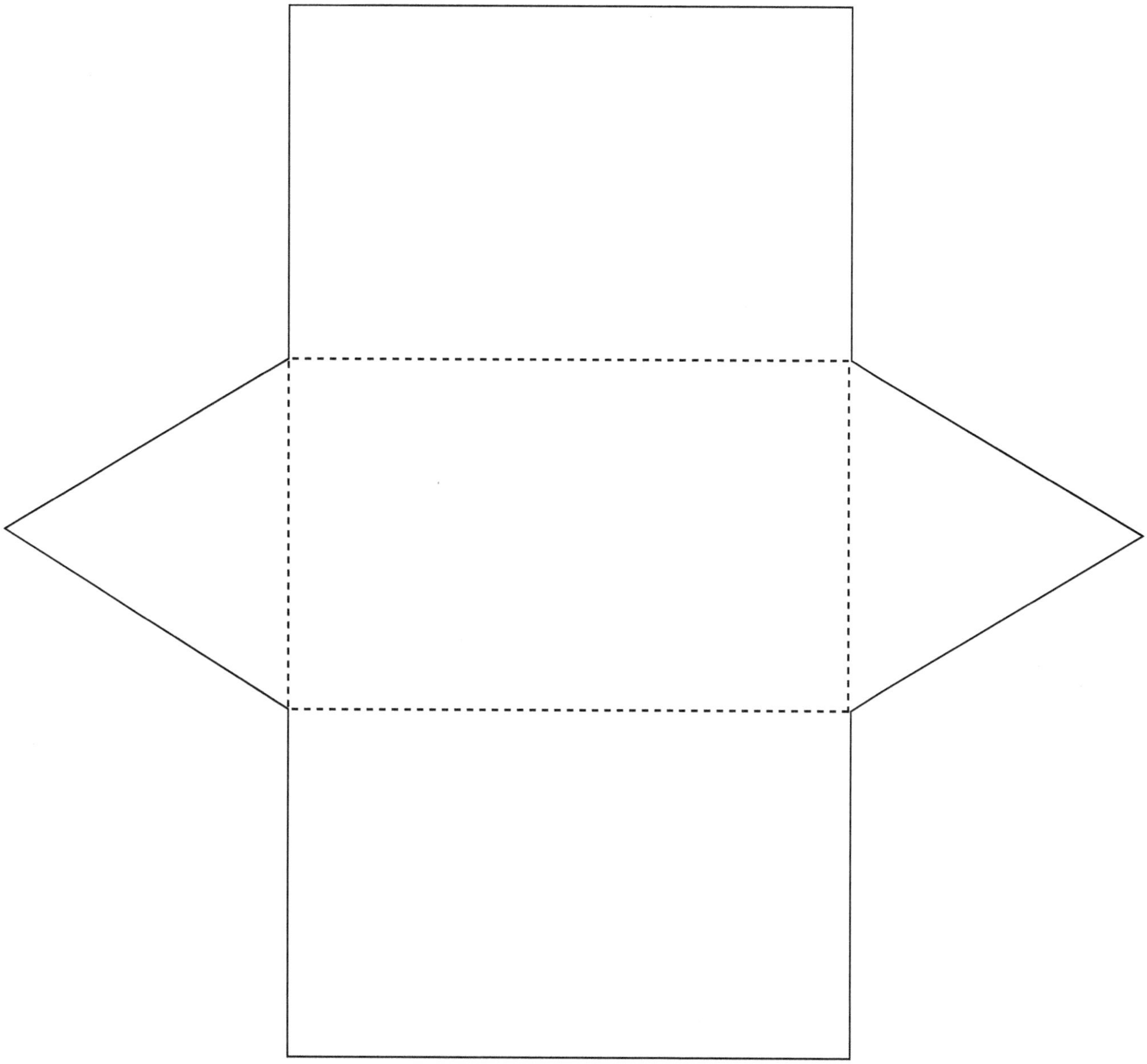

Cut out this net and fold it into its 3D shape.

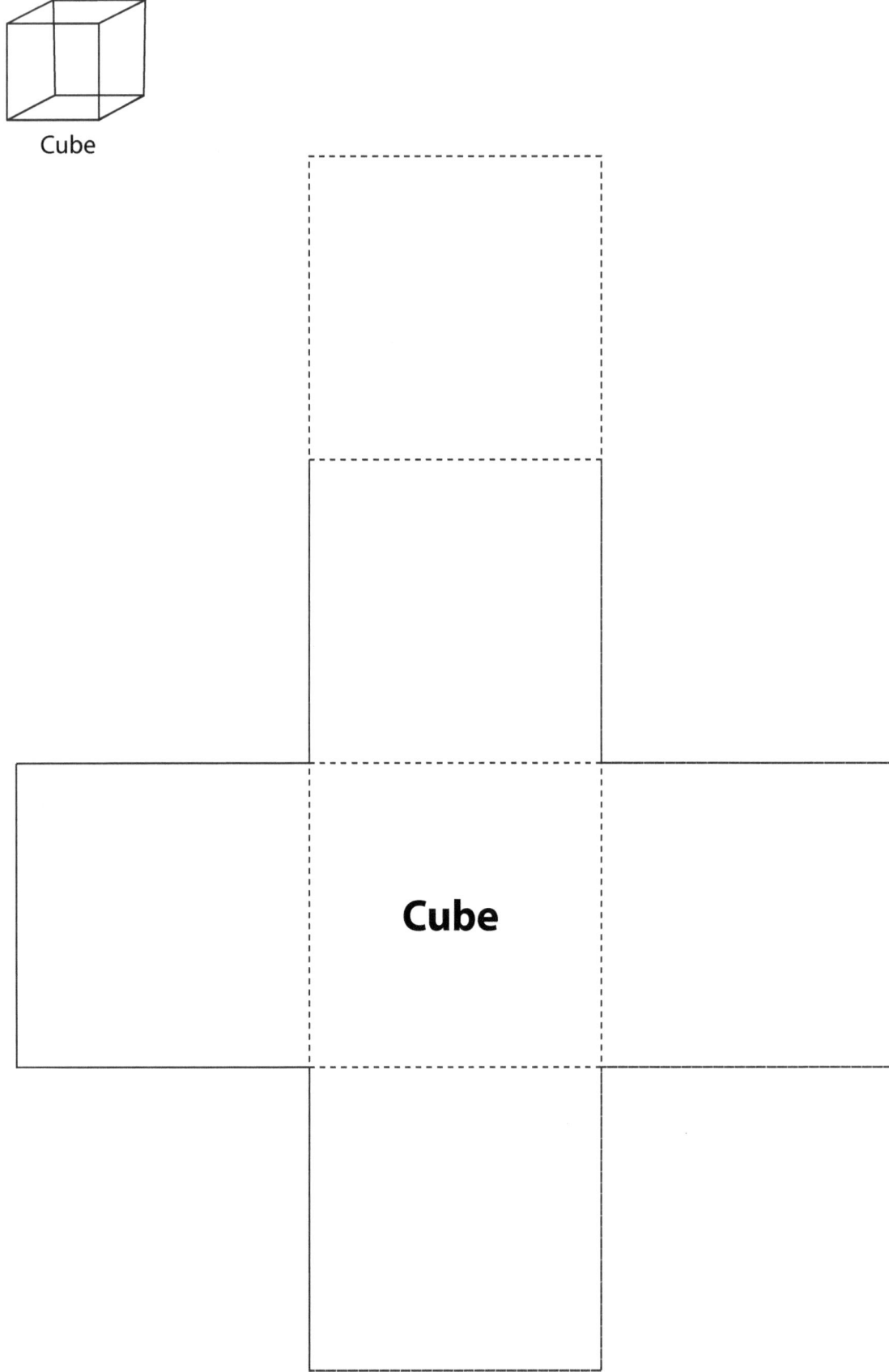

Cube

Cube

Cut out this net and fold it into its 3D shape.

Cone

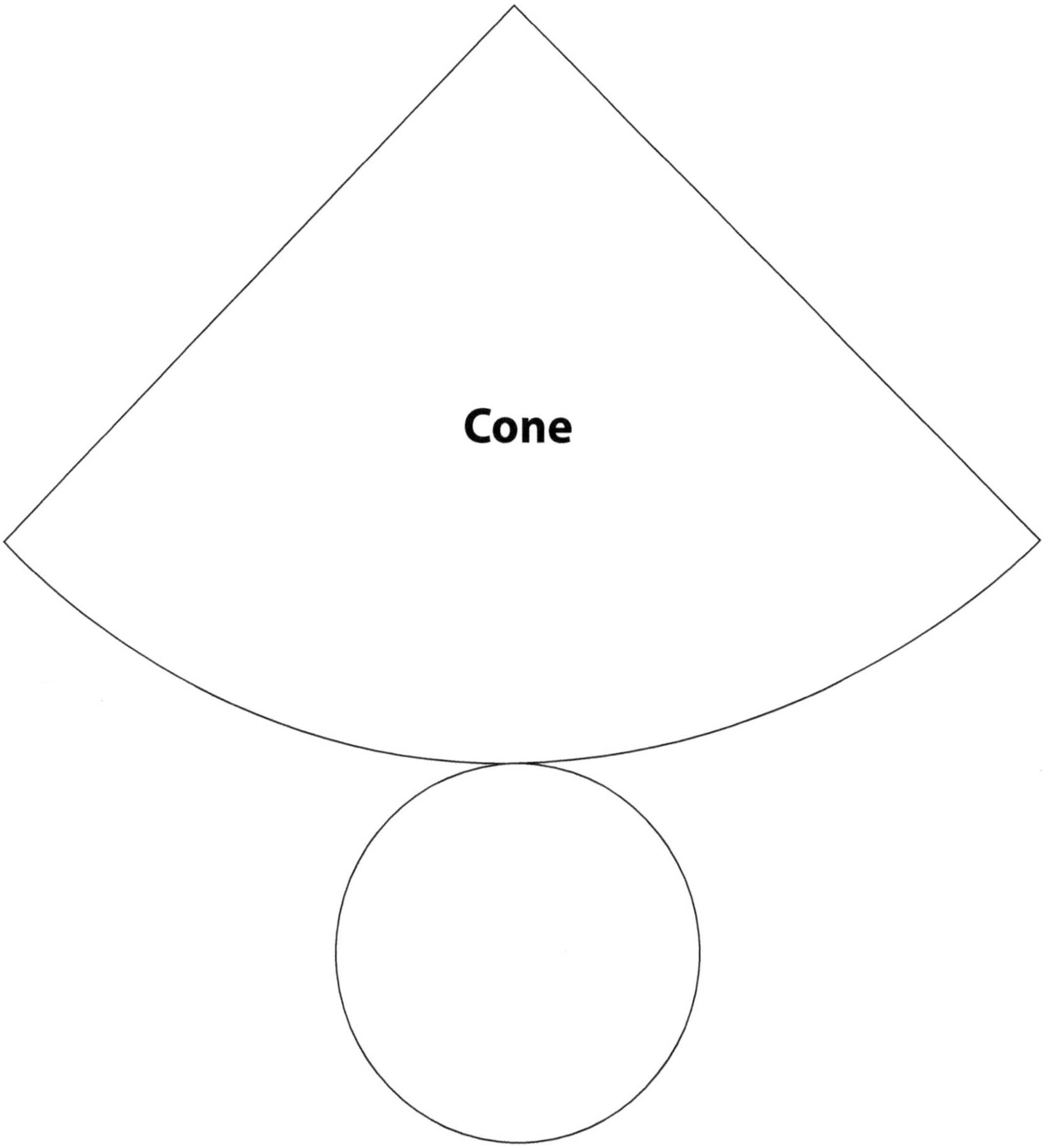

Cone

Cut out this net and fold it into its 3D shape.

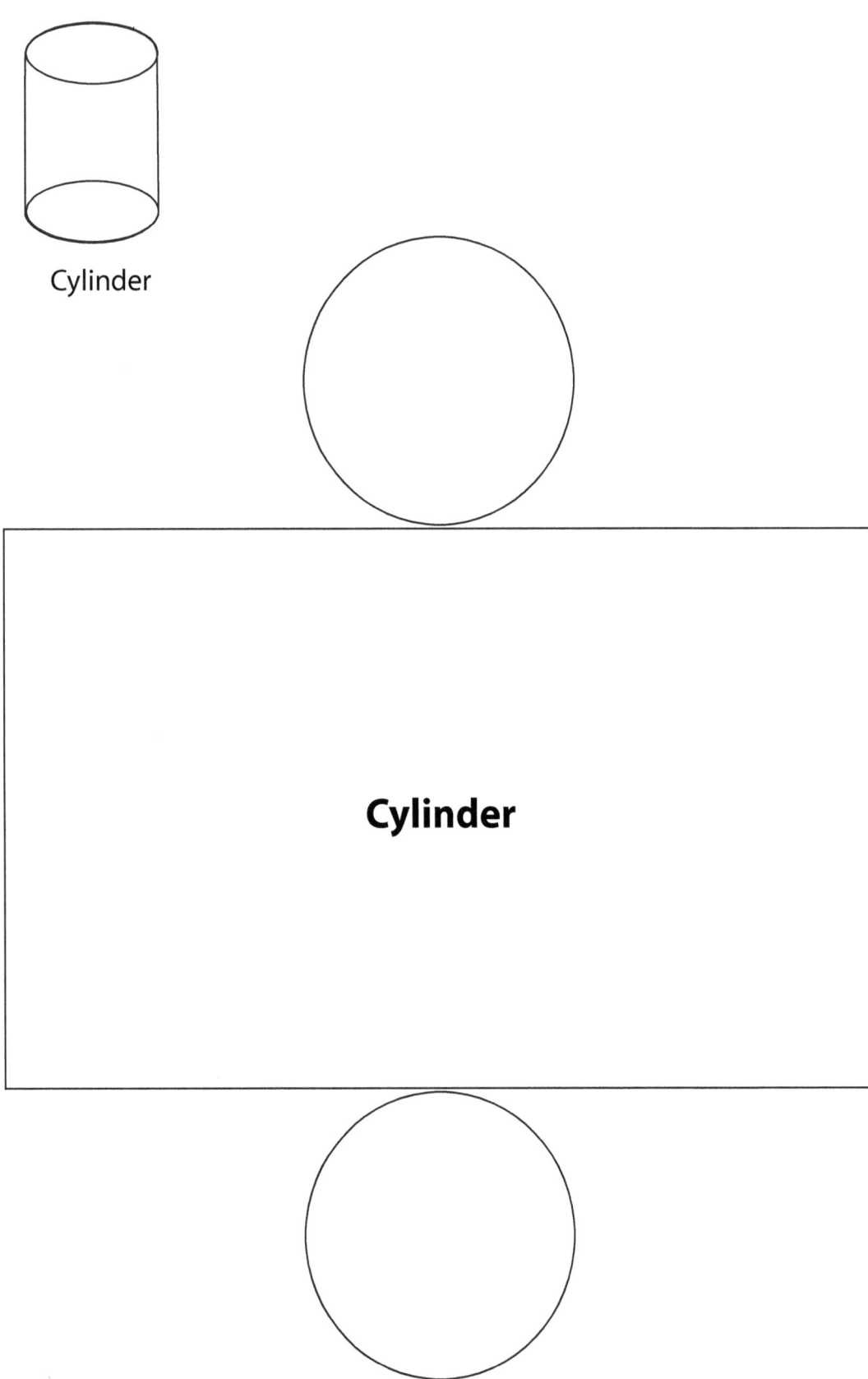

Cylinder

Cylinder

Cut out this net and fold it into its 3D shape.

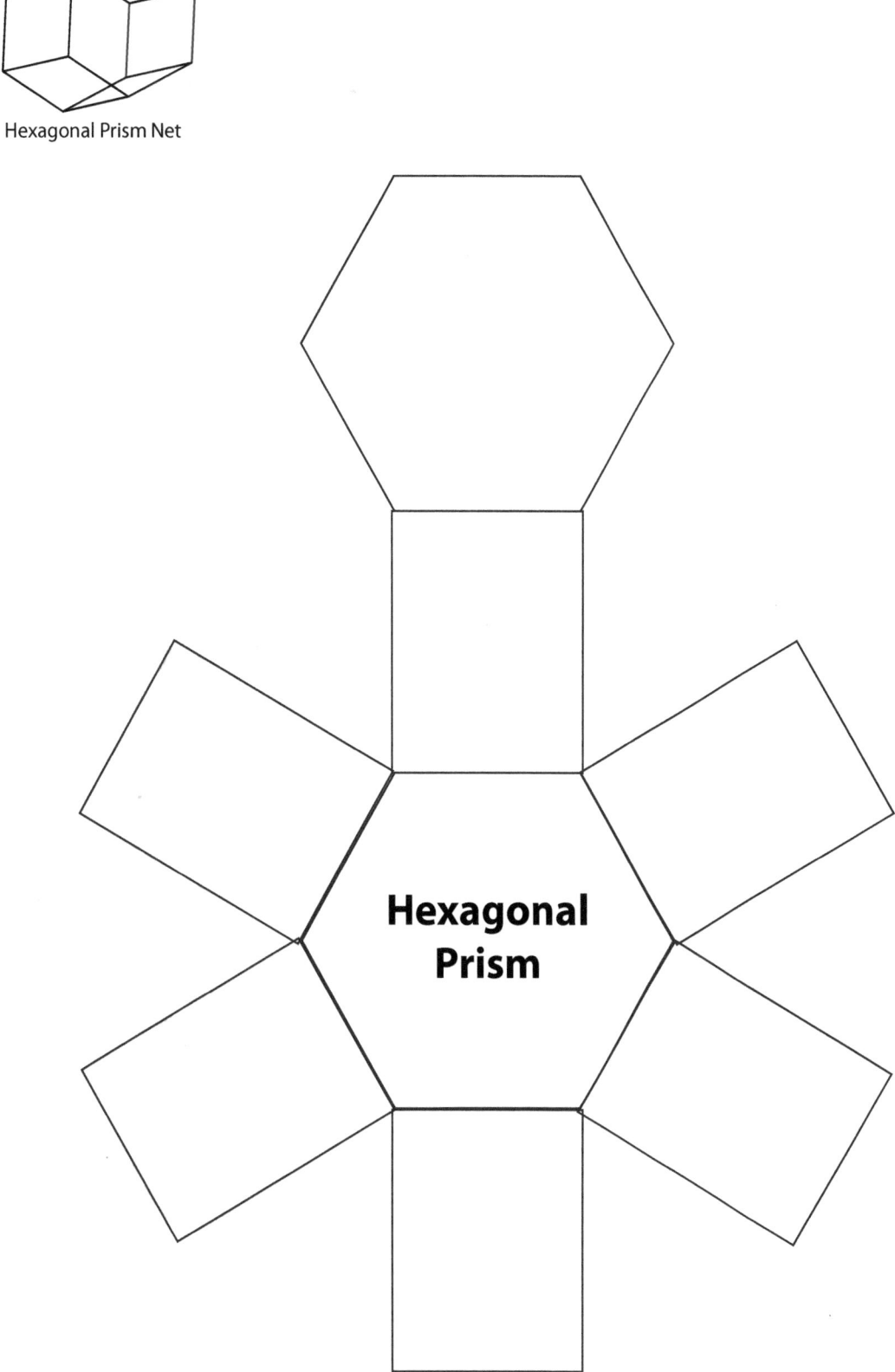

Hexagonal Prism Net

Hexagonal Prism

Cut out this net and fold it into its 3D shape.

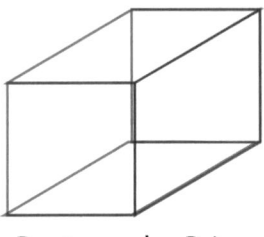

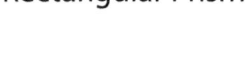

Rectangular Prism

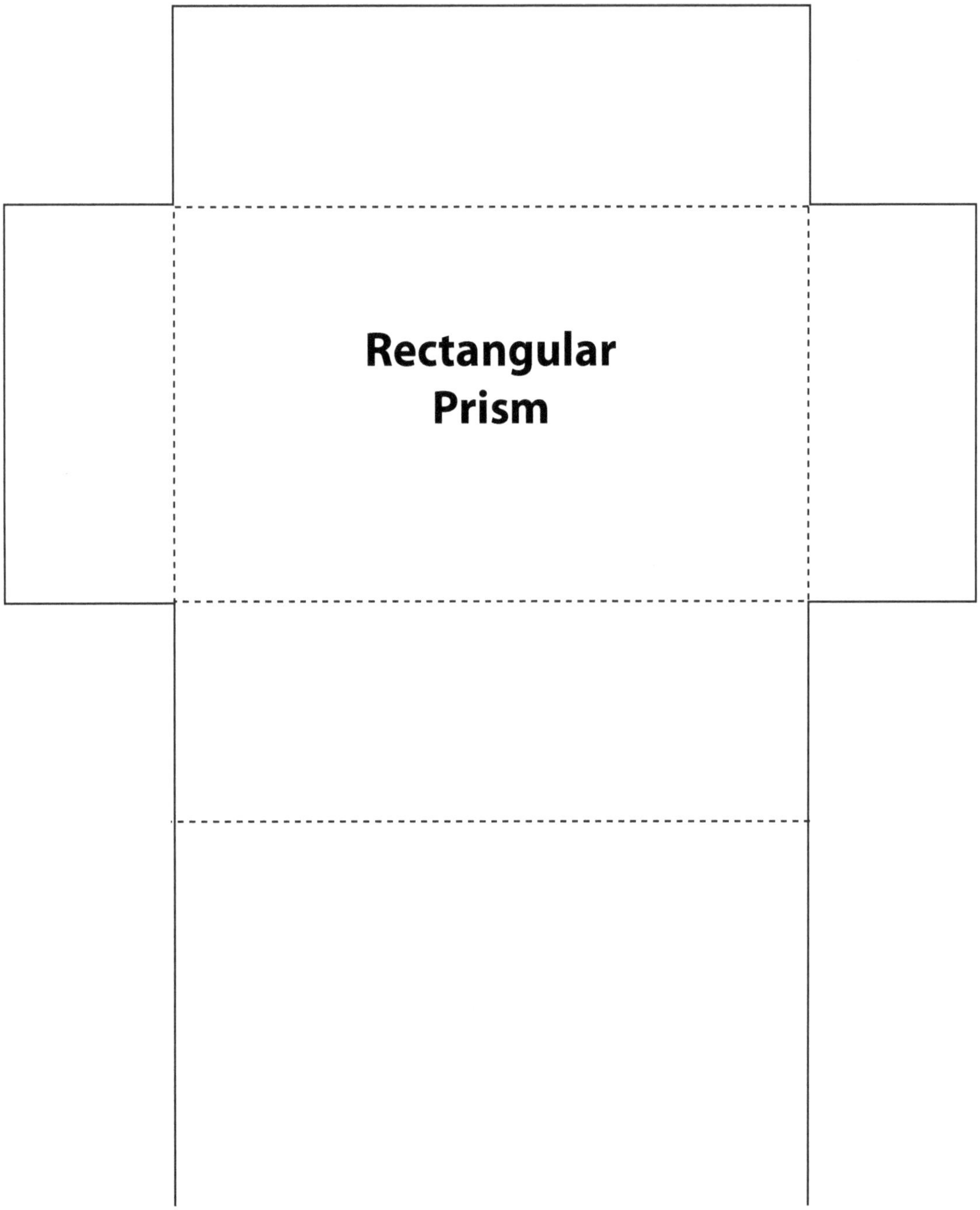

Rectangular Prism

Cut out this net and fold it into its 3D shape.

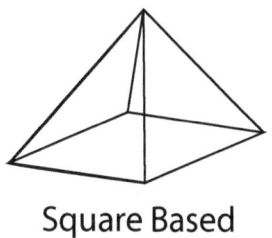

Square Based
Pyramid.

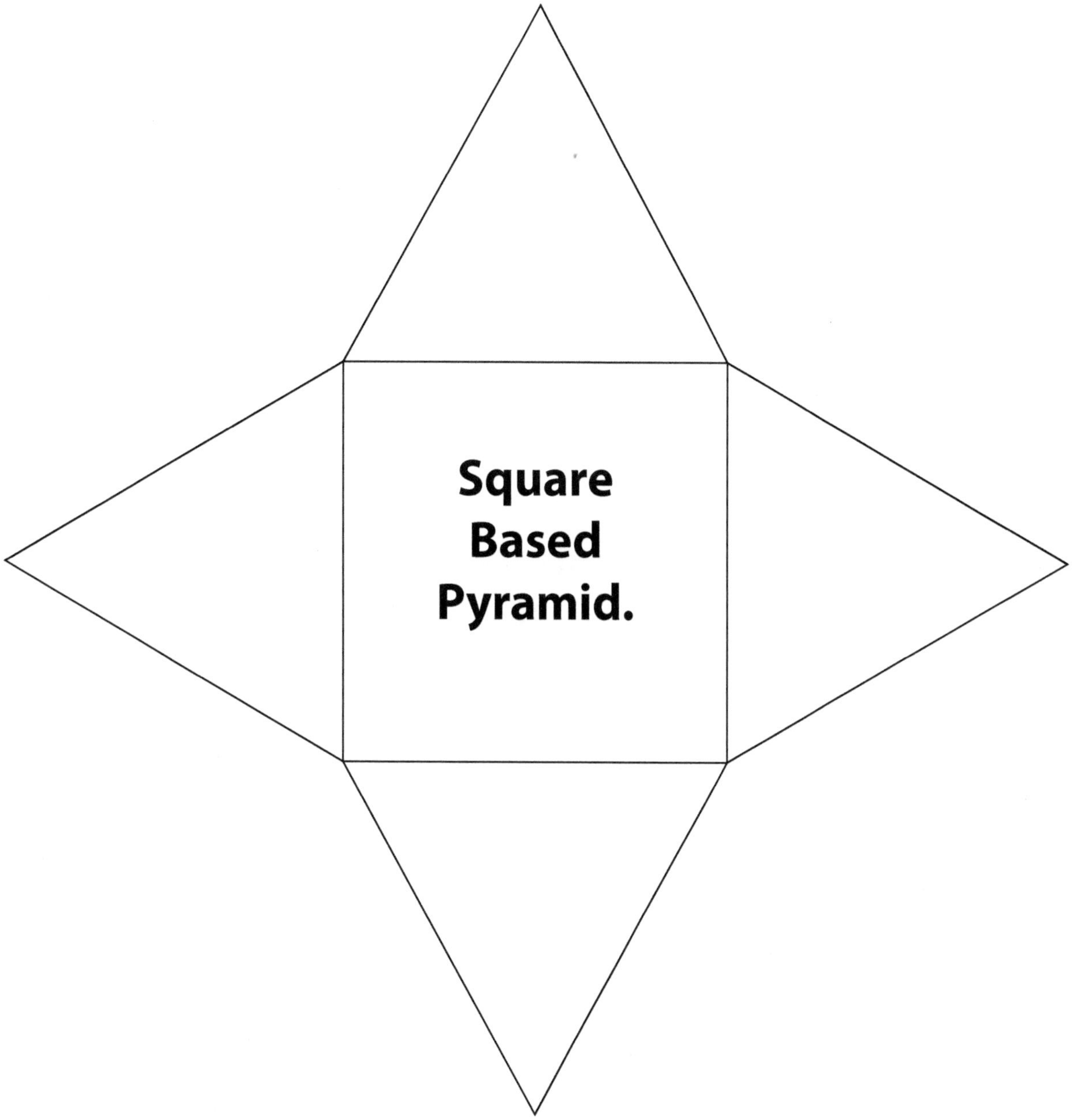

**Square
Based
Pyramid.**

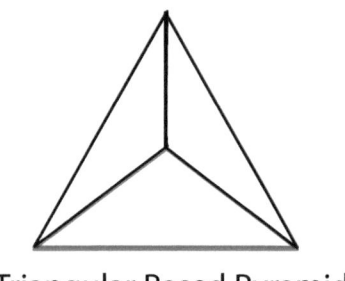

Triangular Based Pyramid

**Triangular
Based Pyramid**

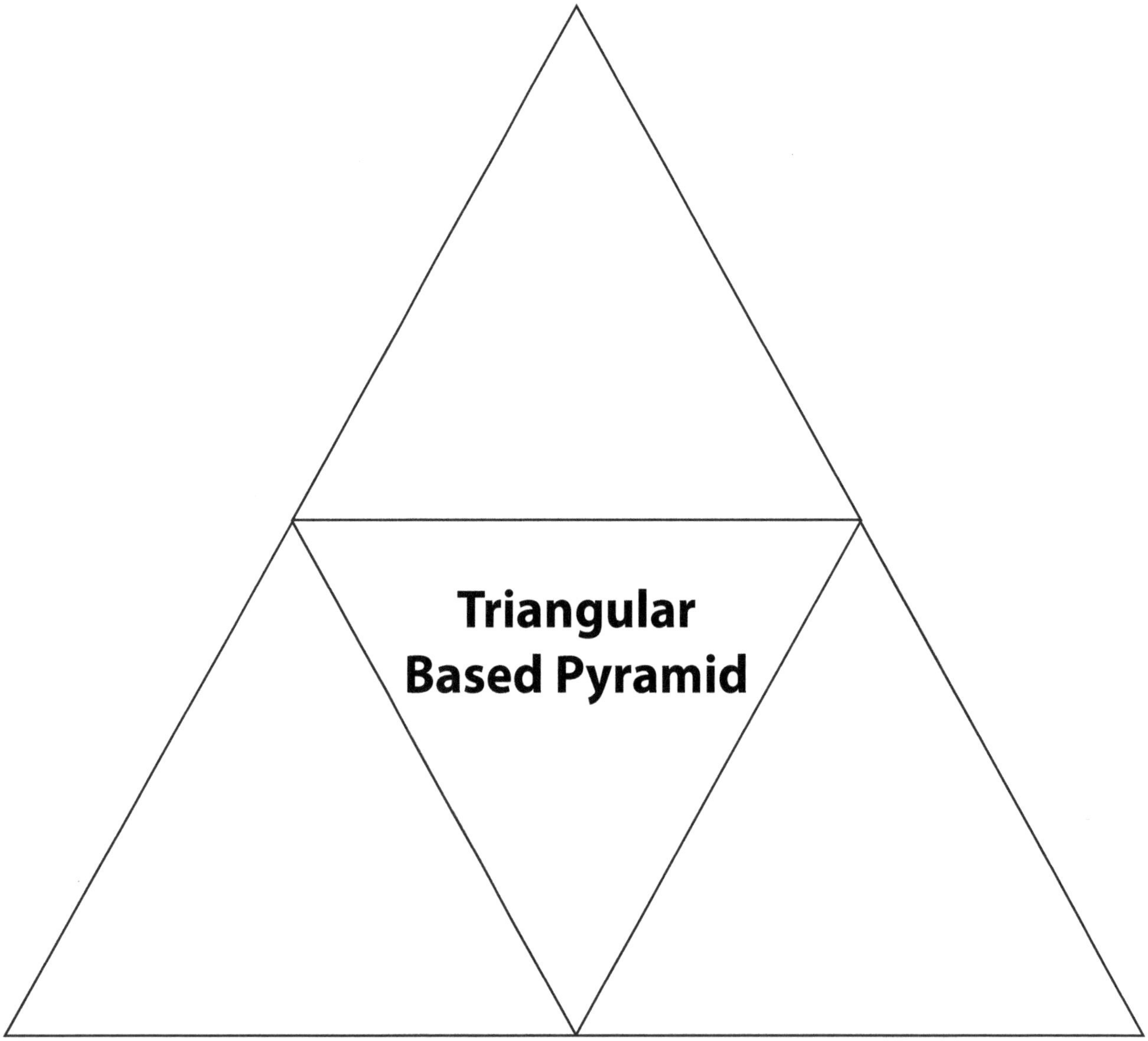

NETS

Match these pictures to their nets

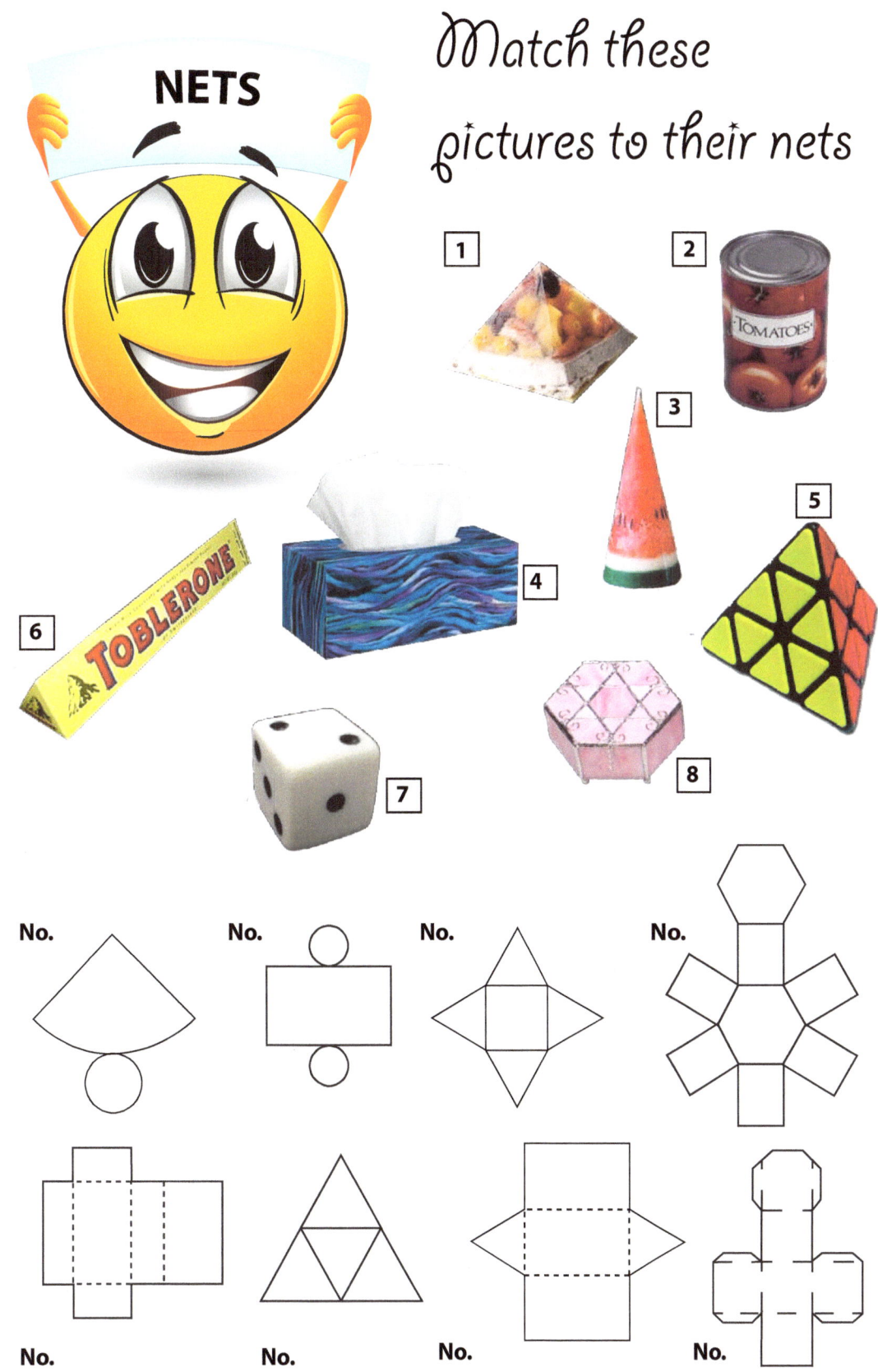

No.

No.

No.

No.

No.

No.

No.

No.